VHDL 101

Everything you need to know to get started

VHDL 101

Everything you need to know to get started

William Kafig

AMSTERDAM • BOSTON • HEIDELBERG • LONDON • NEW YORK
OXFORD • PARIS • SAN DIEGO • SAN FRANCISCO
SINGAPORE • SYDNEY • TOKYO
Newnes is an imprint of Elsevier

Newnes is an imprint of Elsevier
The Boulevard, Langford Lane, Kidlington, Oxford, OX5 1GB
30 Corporate Drive, Suite 400, Burlington, MA 01803, USA

First edition 2011

British Library Cataloguing-in-Publication Data
A catalogue record for this book is available from the British Library

Library of Congress Cataloging-in-Publication Data
A catalog record for this book is available from the Library of Congress

ISBN: 978-1-85617-704-7

Contents

Preface

VHDL, as with so many other languages, is difficult to teach as each part of the language requires the understanding of some other part of the language to make sense. This makes the traditional "linear" approach very difficult as it requires the reader to read the book from start to finish in order to have a sufficient grasp of the language to become useful. This book is experimenting with a different approach – it covers various aspects of the material multiple times! Each "loop" builds upon the previous and introduces new material along with practical application.

The introduction provides a "context" to the reader into the overall architecture of a generic FPGA and the "big picture" of how code becomes a working design. The reader then enters the first "loop" which introduces the reader to the "shape and flavor" of VHDL – rudimentary structures, signals, and a number of things that can be done to a signal in the concurrent (simultaneous) realm. A lab (worked out design that the user can run on his/her own) is provided to show the reader how to "play along" with the book and practice using the various constructs. Specific tools or tool versions are not referenced here as the tools tend to change appearance quite often and explicit instructions may serve only to confuse the reader. A website is provided with the source code and other documents to assist the student in building the reference design used throughout this book.

The second iteration builds upon the understanding gained of simple concurrent statements and delves into one of the most commonly used control structures – the process.

The third and final loop of this book emphasizes code reuse through a construct known as a package. The powerful generic and generate statements as well as functions and procedures are demonstrated to give the reader the tools to create dynamic, flexible "packages" for future use to expedite custom designs. Salient points from the second iteration are driven home so that the reader is fully confident with the material. The final lab leaves the reader with a custom package for future designs.

Within the industry there is a small number of FPGA manufacturers – Xilinx, Altera, Lattice, SiliconBlue, and a few others. As I have first-hand experience designing with the two major manufacturers, Xilinx and Altera, I've seen the good, the bad, and the ugly with both vendors. That said, my strong preference is Xilinx, which conveniently is also the industry leader, so, even though VHDL is vendor agnostic, we have to pick one FPGA vendor as I have neither the time, nor inclination to write the labs to support every synthesis, simulation, and implementation tool currently available.

Xilinx offers a set of free tools (with no time-out or degradation in capabilities), which includes a decent text editor, synthesis engine, and simulator, as well as the implementation tools. This is not a watered down, limited capability, but rather the full-featured tool set. The basic difference between the WebPack (free) and the full price tools is that the full price tools support every current Xilinx device and the WebPack only supports the more commonly selected devices (which is still quite a number of FPGAs and CPLDs!). The other limitation is that the simulation tool (ISIM) limits the total number of lines that can be simulated. We're not getting anywhere near this limit for the labs presented in this book. Please download the latest WebPack for the labs, from www.xilinx.com.

Other things you should know...

This book targets Electrical Engineers wishing to break into FPGA design. It is assumed that the readers have a basic knowledge of digital design and at least a year's experience with the engineering "process," that is, the flow of engineering that takes an idea and turns it into reality.

Why is this book written this way?

1) Many readers have neither the time, nor desire to read through an "academic" stylebook. Often the readers are faced with an immediate need to use or understand VHDL. To meet this perceived need, the book is organized into three sections and an introduction – understanding the FPGA development process with a light introduction to VHDL; commands and data management which covers many of the VHDL language constructs, intrinsic data types, and some of the commonly used IEEE standard data types. This final section focuses on design reuse.
2) This book dovetails with "FPGA 101: Getting Started" which is a "User Guide" on how to successfully develop an FPGA design. "FPGA 101: Getting Started" focuses on the process and Xilinx tools, but not on any specific language. "VHDL 101: Getting Started" elaborates on the language concepts introduced in its sister book.
3) A number of small designs are presented in this book, which serves as the beginning of a "cookbook" of techniques and functions which may be commonly used in many designs. Each design is fully commented and includes a thorough explanation of why certain coding practices were selected.
4) A quick reference guide to both the VHDL language (those items covered in the book) and the common IEEE libraries is provided and cross-referenced to the text of the book.
5) A light, conversational tone is used throughout this book to improve reader comprehension and not to come off as dry as the Mojave in August.

Some familiarity with programming languages is a plus, because, even though this is a book about hardware in an FPGA, how we describe the circuit to the FPGA looks and tastes a lot like programming a microprocessor.

Although simulation is covered in this book, it only scratches the surface. Many simulation concepts, such as self-checking test benches, TEXTIO, and others, are deferred to both existing and future texts and are outside the scope of "beginner".

The bulk of the commonly used VHDL concepts and structures will be covered with enough detail to enable the beginning FPGA designer to accomplish his task, there are; however, certain structures, such as the configuration statement that will also be deferred to a more advanced book.

About the Author

Mr. Kafig escaped from the University of Maryland with a BSEE in 1986 and has been in hiding ever since. Believing that the statute of limitations has been reached, he is climbing out from under his rock where he has been practicing electrical engineering for the last 24 years. During his incarceration in college and immediately afterward, he had the opportunity to work for the Department of the Army at the Harry Diamond Laboratories (now the US Army Research Laboratory) for 16 years working on everything from security systems, to airborne radar systems, to cover tracking technologies, to stuff that really can't be talked about – and of course lots of programmable logic and microprocessors! Concerned that that much government service would leave irreparable psychic scarring, Mr. Kafig moved into the private sector building railway electronics and experiencing design in "real world". Eventually an opening became available as a Field Applications Engineer where he, once again, became infatuated with Xilinx FPGAs.

Family issues brought him to, under the cover of darkness, St. Louis where he consulted on a number of projects from High-Performance Graphics Hardware Compositors, to a Team Leader for a phase-array antenna prototype, to traveling North America teaching the joys of Xilinx Programmable Logic.

Currently Mr. Kafig is employed by Xilinx as a Curriculum Developer, and, since Xilinx prefers him to build FPGAs course content rather than write books, he has taken it upon himself to write this minor tome.

Acknowledgments

My parents – who provided the safe, supportive environment for me to cause my mischief without blowing up too much. And let's not forget that TRS-80 got me started programming! Oh, and I'm sorry about all those partially reassembled things I took apart decades ago. I'll put them back together real soon – promise!

John Speulstra – you earned my sincere appreciation when, those many years ago, you took a young man into the lab and showed him the value of real hands-on training in the tradition of a true mentor – the skills you taught me on the bench have made me a far better engineer and saved the lives of countless pieces of test equipment. Your teaching style and skills have been passed on to many a novice engineer over the years. Whenever I mentor a junior engineer or create educational material – you are there.

CHAPTER 1

Introduction and Background

1.1 VHDL

VHDL. Four little letters. So much to understand.

There are many approaches to learning the Very-High-Speed Integrated Circuit Hardware Description Language (VHDL). These range from the academic – where each piece is looked at, turned upside down and inside out, and fully understood at an atomic level, yet only a few understand how to actually USE the language to implement a design; to OJT[1] where an engineer may be handed either poorly written code (or in rare cases a well written piece of code), or nothing at all and told that they are now responsible for the Field Programmable Gate Array (FPGA) in the system.

Somewhere in between lays a happy medium – a way to learn VHDL and be productive as quickly as possible.

I have a strong background in the Japanese Martial Art of Aikido. Like so many Martial Arts there are "beginning techniques", "intermediate techniques", and "advanced techniques", yet all of them are built on certain fundamental ideas of movement. It would be a disservice to a student to spend hundreds of hours on wrist techniques alone without covering falling techniques (one way to avoid harm), or posture, or any number of other important aspects of the Art.

We start out simply, get a rough idea of what we're in for – sort of a sampling of the various aspects, but none in any great detail.

Now that the student has an idea of what he's in for, he[2] has a better idea of what lies ahead so that the next iteration of falling, arm techniques, wrist techniques, knee work, etc., there is a *context* in which to place the learning. In this fashion, an *upward spiral* of success is achieved.

---

[1] On-The-Job-Training, most often this is a "sink-or-swim" scenario. Without a framework, engineers who learn in this style often code inefficiently.

[2] I use "he" instead of the awkward "he/she" or the more encompassing "he/she/he–she/she–he/it". This is not out of sexism, but out of long tradition, in virtually every language, of using the third person masculine for when the gender of the reader is unknown (as opposed to indeterminate).

VHDL 101.

I'd like to take you on this same path in VHDL. We'll start out by traversing the general process of developing an FPGA.[3] During this process you will encounter "forward references", that is, references to aspects of VHDL that we haven't covered yet. Currently, there is no known way to avoid this problem except through a highly academic, "atomic" view of the language, with no references or context relative to other aspects or commands of the language. This is exactly what we are attempting to AVOID. When you see a new construct, there will be a brief conceptual description of what it does and the details will be provided later.

Once we've made our first "loop" through the material, we'll revisit the initial concepts and statement and begin adding to it and exploring the capabilities of the language. We'll also begin looking at some of the synthesis tool's capabilities and how to verify that the code is behaving properly prior to testing it in silicon (i.e. simulation).

The focus of this book is on VHDL which is one of the primary FPGA languages, but who wants to work in a vacuum? References to various tools and other languages are made to help the reader better understand the landscape into which he is entering. The Xilinx tools will only be covered to the extent that they are required to produce a viable netlist and for simulation. Other books, websites and classes from the FPGA vendors delve into great detail as to how the various tools can be effectively applied.

This book is, by no means, the end-all, be-all reference for VHDL. The intention is to convey the basic principles of VHDL with a smattering of FPGA techniques so that the VHDL has relevance and so that you can get started coding quickly. In that vein neither every aspect of VHDL is covered, nor are the covered aspects detailed to the *n*th degree – but they are covered *enough* to get you coding for FPGAs very quickly.

There are a number of very good, extremely detailed books out there – look for anything written by Peter Ashenden.

1.2 Brief History of VHDL

Way back, in the BEFORE TIME, say around 1981, the US Department of Defense began writing the specifications for the Very-High-Speed Integrated Circuit (VHSIC) program. The intent was to be able to accurately and thoroughly describe the behavior of circuits for documentation, simulation, and (later) synthesis purposes. This initiative was known as VHDL. July of 1983 saw Intermetics, IBM, and Texas Instruments begin developing the first version of VHDL, which was finally released in August of 1985.

[3] Although the focus of this book is VHDL, I believe that a brief overview of the entire flow helps create a context in which to place this learning.

Just to flesh out the timeline, Xilinx invented the first FPGA in 1984, began shipping parts in 1985, and was supporting VHDL shortly thereafter using a third-party synthesis tool.

Due to the popularity of the language, the Institute of Electrical and Electronics Engineers (IEEE) convened a committee to formalize the language. This task was completed in 1987 and the standard became known as IEEE 1076-1987. This standard has been revisited a number of times (1993 and 2000) since then, and the current standard IEEE 1076-2008 is known informally as VHDL-2008. As this latest standard has not yet been ratified (at the time of writing), we will use the IEEE 1076-2000 standard which tends to be a superset of previous versions.

VHDL is a large, complex, and powerful language. Although the language is feature-rich, many designs only require a small subset of the total capabilities.

As a Department of Defense (DoD) project where, quite literally, lives were at stake, no aspect of VHDL would be assumed and it was deigned to be a "strongly typed" language, sharing many concepts and features with the programming language Ada.

VHDL gained very rapid acceptance (as it was mandated by DoD of their agencies and contractors), initially on the East Coast. This is in contrast to Verilog, another FPGA "programming" language, which began its life as a programming language for Application Specific Integrated Circuits (ASICs) and is particularly strong on the West Coast.

VHDL's history still weighs on the current implementation, most notably in the "process sensitivity list" used for simulation. At the time that VHDL was developed, the then "brand new" PC class machines were running at 5 MHz and 640 KB of memory. Processor cycles were very expensive! So, in order not to waste processor time, the process sensitivity list was examined to see if key signals had changed since the last iteration. If so, the block of code was executed, otherwise it was skipped. This is one of the places where some problems of today can occur as the synthesis tools ignore the process sensitivity list and we encounter an error known as "simulation mismatch".

1.2.1 Coding Styles: Structural vs. Behavioral vs. RTL

You may hear several adjectives used with VHDL – Structural VHDL, Behavioral VHDL, RTL coding, etc. These are not separate languages, but rather different approaches to coding. These methods are not mutually exclusive, but may be mixed freely within a project or even a single module.

Structural VHDL or structural coding is a very low-level style of coding. This practice involves instantiating, or calling out explicitly, specific instances of primitives or previously defined modules and wiring them together. Although an entire system can be developed this way, in practice this is rare as this method is extremely time-consuming. There are three places where this style is typically used – at the top level of a design where large lower-level modules are stitched together, when a specialized piece of Intellectual Property (IP) is provided often as a "black box"[4], or when specific architectural features such as block memory, high-performance communication cores, or other specialized silicon resources, are required.

RTL stands for Register Transfer Level and refers to a coding style that defines how signals flow between hardware registers and the logical operations that occur between the registers. This is one step more abstracted than structural coding as the independent logic elements don't need to be instantiated but can rather be inferred[5] from equations.

The next level more abstracted is called "Behavioral Coding". This adds an additional layer of abstraction in that software-type constructs such as "for loops", "if statements", and "case statements". There is a common misunderstanding that "Behavioral Coding" can only be tied to simulation – this is most certainly not the case! While it is true that some Behavioral Coding constructs cannot be synthesized, many can be. This term is most often confused with "Behavioral *Modeling*" which is reserved for simulation.

Since simulation test benches run only on the development platform and aren't required to generate hardware they may contain constructs which cannot be translated into a hardware structure. One may accurately consider VHDL in the simulation domain, a superset of capabilities over VHDL for synthesis (conversion into hardware). Whether you choose to code in the Behavioral, RTL, or Structural style has nothing to do (other than the use of non-synthesizable constructs) with whether code is strictly synthesizable or not.

1.3 FPGA Architecture

Coding effectively, regardless of the language you select, requires some basic understanding of the device's architecture, so we'll take this opportunity to have a very brief overview of the main aspects of a generic FPGA's innards (see Figure 1.1[6]).

Notice the columnar organization of this device – Xilinx calls this "ASMBL" for Advanced Silicon Modular Block. This enables Xilinx to rapidly (and reliably) construct multiple variations within a family by adding additional columns and/or blocks of rows to create a device to meet customer demands.

---

[4] A black box, in this context, is a piece of code, usually provided as a netlist. The vendor will provide a description of the signals entering and exiting the module which is all the information required to instantiate this black box.

[5] Inference indicates that the synthesis tools make certain decisions based on user settings as to how to convert the described function into FPGA primitives. The user may provide an abstract concept, such as addition, and the synthesis tools will decide how many multiplexers, LUTs (combinatorial logic generator), or other primitive resources are required to implement the specified equation.

[6] This overview does not represent any specific FPGA, rather a composite of many of the features typically found in a modern FPGA.

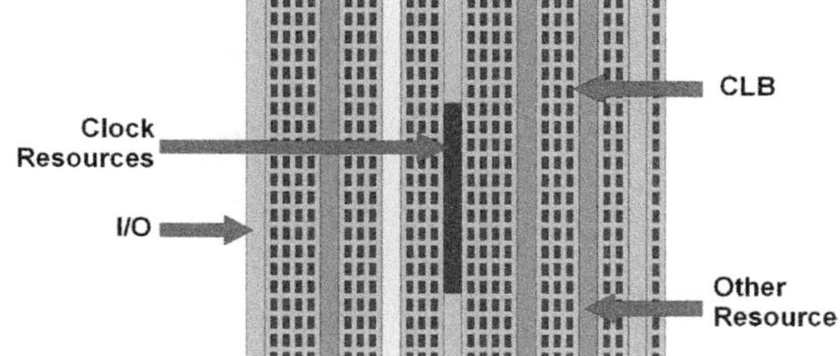

Figure 1.1: Coarse Look at a Generic FPGA Architecture.[6]

The columns on the leftmost, rightmost, and center represent I/O resources. An I/O resource may consist of any mix of input and output buffers (many support both single ended and differential), registers for supporting Double Data Rate (DDR), programmable termination resistors, programmable delay elements, and serializer/deserializer modules. These I/O blocks effectively connect the package pins to specific signals internal to the FPGA.

The short dark region in the center column represents some of the clocking resources. These resources include devices for multiplying up (or down) the incoming clock signals using Digital Clock Managers (DCMs) and/or analog Phase Locked Loops (PLLs). There are also a number of specialized differential clock distribution buffers located in this area.

Other resources such as blocks of memory, multipliers, Digital Signal Processing (DSP) blocks, microprocessors, high-speed serial I/O, PCIe, and other specialized devices may be found in the other columns.

As a general rule of thumb: if you want your design to run as fast as possible with the least impact on power and area, use the dedicated silicon resources wherever possible. If, on the other hand, you must make your code portable to other FPGAs which may lack these resources, you should target the fabric.

Further discussion of these specific silicon resources is out of the scope of this book. Contact your FPGA manufacturer for appropriate User Guides and courses.

The small blocks interspersed among the columns are, in the Xilinx lexicon, referred to as Configurable Logic Blocks (CLBs). These blocks support both combinatorial logic processes as well as registers. These CLBs are comprised of more granular modules called "Slices". Depending on the family, there may be two or more Slices per CLB, and each Slice may contain two or more pairs of Look-Up Tables (LUTs) and registers. In addition to the LUT and register, there are multiplexers that are used for combining and directing where the signals go. Generally, this is where the bulk of a design winds up (see Figure 1.2).

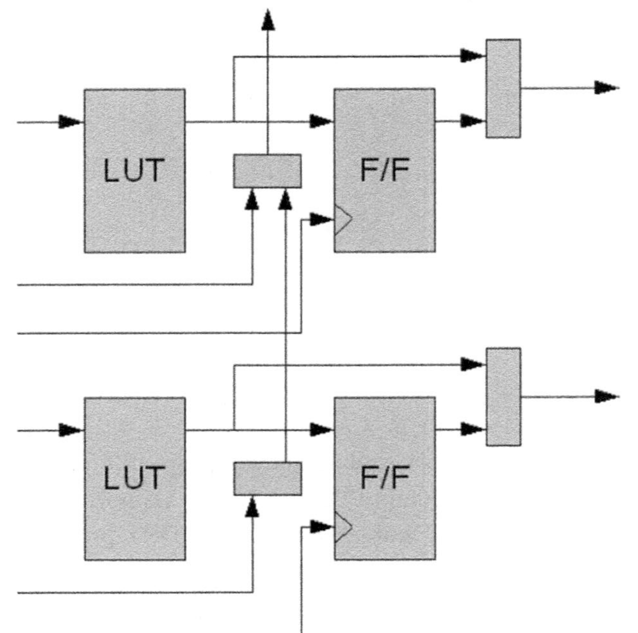

Figure 1.2: Generic "Slice".

These logic and register resources are referred to as the FPGA "fabric" to differentiate it from dedicated silicon resources.

The last major resource, which is not shown in the illustration for purposes of clarity, is the routing which, although surprising at first, accounts for the majority of the silicon area in the device. According to the June 2006 issue of the *IEEE Spectrum*, a typical FPGA contains many kilometers of routing!

You will see numerous references to a "silicon resource". For clarity, this is any single or collective entity within the FPGA whether it is a humble Look-Up Table or LUT (found within a CLB) or a PCIe core or processor.

All of the interconnects, FPGA fabric, and dedicated silicon resources are connected via a series of run-time static switches which regulate the flow and connection of information through the system. It is these switches and connections as well as the "equations" programmed into the LUTs that must be focused upon when the code is written.

With that, let's begin with a brief introduction to the overall flow of developing an FPGA and where VHDL fits in this process...

1.3.1 Creating the Design

Volumes have been written on how to successfully execute a design. Since this is a book on VHDL and not on design philosophies, we will not delve too deeply into this topic.

Suffice it to say that design philosophies range from the incredibly detailed "think everything out in painful detail from every angle before you lift a finger to writing a single line of code" (which, by the way, is an excellent way to go for larger designs), to completely ad hoc "sit down and start coding and resolve issues along the way" which only works effectively for small designs.

Organizing the design and resolving issues before coding is where the real designing takes place. Using detail oriented processes, such as those outlined by the IEEE for software development, may be a bit tedious and cumbersome, but results in a higher quality of code which requires little debugging time and, although there isn't much apparent forward progress when writing the design description, it will greatly accelerate the project once the coding has begun.

Many designers choose a point somewhere between these two extremes which consist of a system-level block diagram and some research on specific algorithms to be implemented.

All this said, good design practices are important for a successful FPGA design, so hints, coding examples, and best practices notes are scattered throughout this book.

Regardless of the language or the complexity of the design, the designer must know what he is doing and how it fits into the bigger picture.

Here are some basic points to follow:

1. "Divide and Conquer" – break the design into manageable pieces
   a. Divide by functionality
   b. Complex modules can be further broken into smaller pieces[7] until basic functionality is reached
   c. Write a single sentence describing what each module at each level of hierarchy does. If you can't do this, you don't have a clear enough idea of what the module should do. Think about it some more
2. Draw a top-level block diagram for each hierarchical level in the design
   a. Draw lower levels of the design hierarchy for more complex blocks
3. List the signals which move from module to module
   a. For the same level of hierarchy
   b. Through vertical levels of hierarchy

[7] This is the beginning of hierarchical design.

  c. Don't just list signal names, but include types and units[8]
  d. Choose meaningful signal names – be descriptive, complete, and unambiguous
4. Don't start coding until the above steps are complete

Paper exercise

Consider building a simple clock which shows the date and time. Assume that somebody else is working on the task of setting the clock and that you are responsible only for the tracking and display of time and date.

Use the design process above to create the outline of this design.

Begin by breaking the task of "tracking and display of time and date" into the following modules Track Time/Date and Display Time/Date.

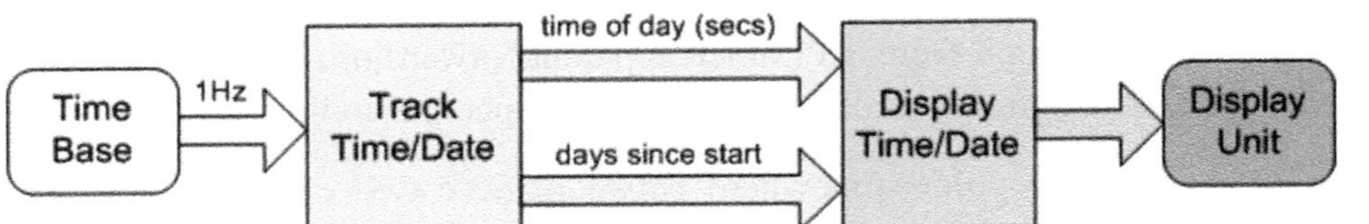

Figure 1.3

Now describe the gross behavior of each module in one sentence:

Time Base – This module produces a 1-Hz pulse, which forms the basis of time in this design.

Track Time/Date – This module counts the number of 1-Hz pulses and produces a count of the number of seconds within the day and a count of days since 1 January 2010.

Display Time/Date – This module displays the information provided by the Track Time/Date module as HH:MM:SS and DD/MM/YY.

Display Unit – This module is responsible for physically displaying the information from the Display Time/Date module.

The above descriptions mention what is done in each module, but further refinement is required to identify the types of signals that pass from one module to the next.

Add signal specifications to each module in one or two sentences.

Time Base – This module produces a 1-Hz pulse, which forms the basis of time in this design. This module is implemented external to the FPGA and produces a 50% duty cycle square wave at 1 Hz ±3 ppm[9]

[8] If you don't think that this is important, I'll refer you to a certain recent Mars mission that failed because some designers worked in metric while others worked with the English system. The net result was not a soft touch down, but a crater.

[9] In a real-life design, this would likely be a 32768 Hz (or faster) oscillator with a divider built into the FPGA.

Track Time/Date – This module counts the number of 1-Hz pulses and produces a count of the number of seconds within the day and a count of days since 1 January 2010. For simplicity, there is no need to track leap years or time shifts due to switching between Daylight Saving Time and Standard Time. Time is passed in seconds as an unsigned integer ranging between 0 (midnight) and 86,399 (1 s before midnight). Days since 1 January 2010 is passed in days as an unsigned integer ranging between 0 (1 January 2010) and 8191 (approximately 22.5 years).

Display Time/Date – This module displays the information provided by the Track Time/Date module as HH:MM:SS and DD/MM/YY. Information is provided to the Display Unit as a series of seven-segment LED std_logic_vectors[10] supporting 12 digits for a total of 84 segments.

Display Unit – This module is responsible for physically displaying the information from the Display Time/Date module. It is implemented external to the FPGA and is comprised of 12 seven-segment LED displays.

Now that each module is understood, modules requiring further clarification can be further subdivided into a more detailed block diagram.

Time Base is most likely implemented external to the FPGA and no further refinement is required.

Track Time/Date can be simplified into two modules, each producing one of the specific types of output from this module.

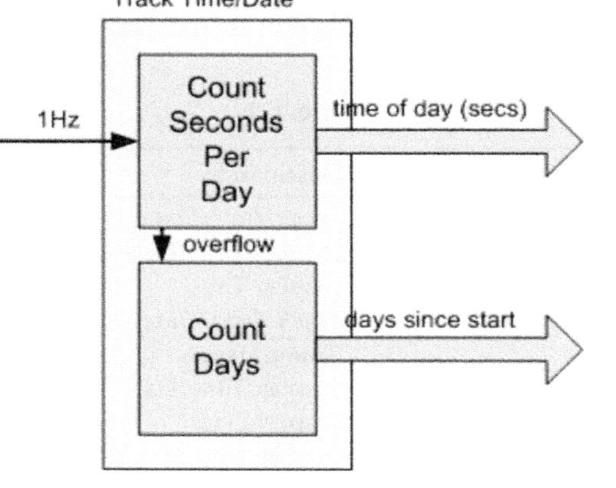

Figure 1.4

[10] A std_logic_vector is a type of signal that will be discussed when data types are introduced. For now, just consider them as binary values.

Count Seconds Per Day produces an integer between 0 representing midnight and 86,399 representing 1 s before midnight. Upon reaching the terminal count of 86,399, this module will reset the count to 0 and simultaneously issue an overflow to the Count Days module.

Count Days module increments by one on the first second of each day based on the assertion of the overflow signal from the Count Seconds Per Day module. The count provided by this module ranges from 0 to 8191. Upon reaching the terminal count of 8191, the count will reset to 0 with the understanding that the date is no longer valid.

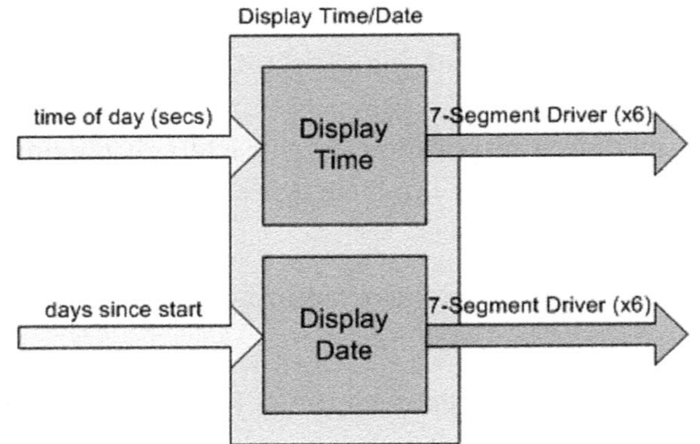

Figure 1.5

Display Unit is implemented external to the FPGA and no further refinement is required.

Next, name the signals flowing from one module to the next:

Table 1.1

Module	Signal Name	Destination	Description
Time Base	clk1Hz	Track Time/Date	1-Hz clock
Track Time/Date Count Seconds Per Day	time_of_day	Display Time/Date Display Time	Integer 0–86,399
Track Time/Date Count Seconds per Day	overflow	Track Time/Date Count Days	std_logic (1 bit)
Track Time/Date Count Days	days_since_start	Display Time/Date Display Date	Integer 0–8191
Display Time/Date Display Time	LEDs_time	Display Unit.time segments	std_logic_vector (42 bits)
Display Time/Date Display Date	LEDs_date	Display Unit.date segments	std_logic_vector (42 bits)

At this point we have enough information to build the outline of the design which we will explore once we start into the VHDL section of this book.

CHAPTER 2

Overview of the Process of Implementing an FPGA Design

There's more to designing an FPGA than just typing in the VHDL program. An entire process exists for turning what you just typed into an image suitable for loading into an FPGA. The code that you enter determines both the *behavior* and *performance* of the design.

Specifying the *performance* standards, that is, how fast the design should operate, is an art in and of itself and is well outside the scope of this text.[1] These performance standards are specified in a User Constraint File (UCF). Suffice it to say that the code that you write determines what gets created[2] – that is, how it behaves,[3] but how it is placed within the fabric of an FPGA is determined by the implementation phase and is guided by performance and location constraints (see Figure 2.1).

2.1 Design Entry

The process begins by entering your design into the tools.[4] The most common method is by using one of the two primary Hardware Description Languages (HDLs), which is probably why you bought this book on VHDL in the first place. Most modern FPGA development tools support multiple languages within a project, so you can mix VHDL with other design entry methods if you choose. The flow diagram in Figure 2.1 is illustrative of the typical flow through the Xilinx tools. The tools pick up at the design entry stage with a decent editor. Please refer to Figure 2.1 as you work your way through the explanation of the stages that follows.

1 The various FPGA vendors provide differing levels of training as to how to derive the highest level of performance from your design. Xilinx offers both free on-line recorded e-learning modules as well as instructor-led classes (not free, and worth every penny).

2 What you code isn't *EXACTLY* what gets created – the synthesis tools will likely optimize your code based on synthesis constraints and tool options.

3 While the UCF specifies how fast the design *should* operate, it doesn't alter the design to run at that speed. The way that you code will have a significant impact on the overall design performance.

4 In actuality, entering your design should be toward the end of the design process. As was mentioned in Chapter 1 the first step is writing appropriate specifications for the design followed closely by constructing an appropriate design.

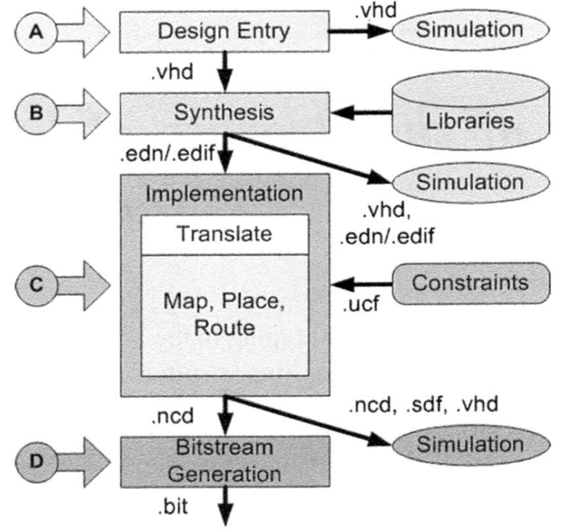

Figure 2.1: Typical Xilinx FPGA Development Flow.

Another very popular hardware description language is Verilog. Since the mention of Verilog has reared its ugly head let's take a moment to mention that there is almost a religious war between VHDL and Verilog. If you want to learn more about the differences between these two languages, see Sidebar 2.1.

Looking forward, there are a number of other HDLs on the horizon such as SystemC and System Verilog. These languages are beginning to gain some traction; however, at the time of writing, these other HDLs represent only a small portion of the total HDL market.

Often development tools vendors will provide additional entry mechanisms, typically graphically based, such as schematic capture and state machine diagram capture. Frequently, these tools generate VHDL or Verilog on the back end for the synthesizer.

Specialized processes, such as embedded processor designs and Digital Signal Processing (DSP) development, typically employ very sophisticated graphical interfaces to collect information about the design and can translate the graphics into an HDL for later synthesis or generate a netlist directly. Xilinx's Embedded Development Kit (EDK) and Matlab's Simulink are examples of such tools.

All you really need for design entry is a good text editor. There is a wide number of free text editors available, many of them with language context coloring (i.e., keywords appear in a different color than signals and variables, comments appear in yet another color, etc.). Xilinx's free Webpack offers a fairly credible text editor and is integrated into the development environment.

SIDEBAR 2.1

VHDL vs. Verilog – Which One Is Better?

Years ago when HDLs were a very new concept, there were two camps – one backed by Department of Defense (DoD) (VHDL) and the other by the ASIC developers (Verilog). VHDL was originally designed as a means of documenting hardware and simulation – bear in mind that the only programmable logic available at this time were small PLDs (Programmable Logic Device) such as the 22V10 and these were usually programmed in an existing language – ABEL (Advanced Boolean Expression Language).

Since projects for the DoD usually have life and death implications, the DoD prefers to use "strongly typed" languages which, by their nature, are somewhat more verbose, but tend to avoid certain types of mistakes. This leads many people to speculate (incorrectly) that Verilog is easier to learn than VHDL.

During the early days of HDLs, VHDL sported a number of advantages over Verilog in that there were a number of constructs that were simply not available in Verilog. Over the course of the last 15 or so years there have been a number of "upgrades" to both VHDL and Verilog which make the two languages nearly identical in capability.

Theoretically, the same design implemented in VHDL and Verilog should produce identical net-lists, therefore, they should have the same performance. This, however, is rarely the case. Verilog tends to have fewer lines of code per module simply because it is a "loosely typed" language, that is, signals and variables (wires and variables in Verilog) don't have to be defined prior to their use as is the case with VHDL, which occasionally leads to errors which have to be ferreted out later (usually by reading the detailed synthesis report) rather than being flagged up front during synthesis.

Verilog also supports a different set of "scoping" rules – although in many instances convenient, it runs the risk of generating "side-effects" and may make the code more difficult to maintain although it might have been easier to write.

2.2 Synthesis

Once the design has been entered, it must be synthesized. The process of synthesis involves converting the VHDL source files into a netlist. A netlist is simply a list of logical elements (things that combine, change, or store digital signals) and a list of connections describing how these elements are wired together. A netlist *can* be platform independent, that is, it can target any architecture from any vendor. As we will see later, making use of the specialized resources within the FPGA is almost always the better choice as these specialized resources consume less power, fewer general purpose resources, and meet timing (performance goals) more easily.

All errors generated by the synthesis tools must be resolved prior to moving to the next step. Most synthesis tools will also provide a copious list of warnings. Many of these warnings and informational messages will simply echo what you deliberately and intentionally specified;

however, these messages should not be categorically ignored as they describe a number of "issues" that the synthesis tool has with your code and may reveal mistakes or unintended behaviors in your code.

Many development environments provide additional support tools such as Register Transfer Level (RTL) viewers (graphical representations of the source code) and/or netlist viewers (graphical representation of how the source code will be implemented in fabric). These tools can be very helpful in tracking down certain types of coding issues.

The output of this stage of the process is one or more netlists. These netlists come in one of several flavors, the most common being the EDIF/EDF/SEDIF and EDN formats. The Electronic Design Interchange Format (EDIF) is a more widely accepted format. The output of the Xilinx synthesis tools is a ngd file which stands for Netlist Generic Database and conveys the same information as EDIF, but in a binary (not ASCII) fashion. Xilinx provides tools to convert NGD to EDIF should a manual review of the netlist be necessary.

Generally, reviewing the netlist is an unnecessary step which is only performed when there is sufficient reason to believe that there is an issue with the results of the synthesis tools.

2.3 Simulation

Simulation is all too often overlooked during the development of an FPGA. This usually leads to unnecessarily long verification and debugging times.

Best design practices call out that as each module is completed, it should be simulated. When a design is properly executed, it is conceptually created from the top-down, but constructed from the bottom-up. Generally, the deeper in the hierarchy of the design a module is, the easier it is to simulate and verify. With the confidence that the building blocks are working as intended and as more lower-level modules are completed, they can be combined in a higher-level module with a stronger confidence of behaving properly.

Simulation requires the following:

- the design
  - as source code, or
  - as a netlist, or
  - as a complete implementation
- and some type of stimulus
  - usually a test bench, but may also be
  - a Tcl script or other type of script which provides signal stimulus

A test bench is one or more modules that connect your design, the Unit-Under-Test (UUT), with internally generated stimulus or stimulus from a file to drive the inputs of the UUT and

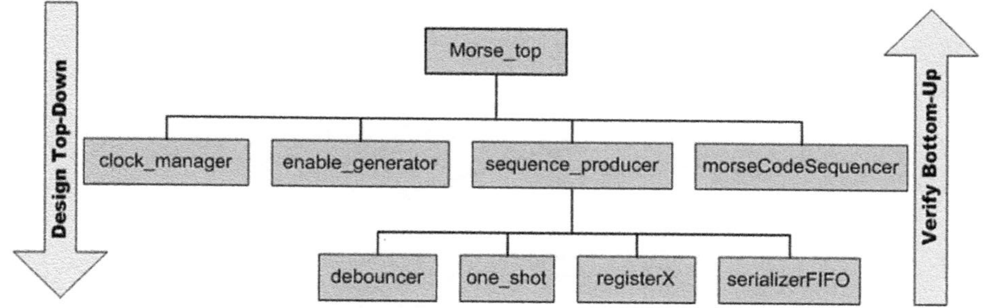

Figure 2.2: Example Design Hierarchy – Design From the Top-Down, Build and Verify from the Bottom-Up.

may collect and process the outputs of the UUT. Ideally all possible sets of stimulus should be included to fully validate the design. We will be writing simple test benches throughout this book to reinforce the best design practice of simulate small, simulate frequently.

After each module is written, it should be simulated. Usually the low-level modes are relatively simple and the test benches quick and easy to write. Once simulated the designer has confidence to move to the next module so that when the lower-level modules are combined, the focus is on the verification of the combination and additional logic, rather than on debugging the lower-level modules (see Figure 2.2).

There are four primary locations for running simulation (Figure 2.3):

A. Behavioral Simulation – prior to synthesis
B. Netlist Simulation – post-synthesis
C. Post-Map Simulation
D. Post-Implementation or Post-Place-and-Route Simulation

A. Simulation run prior to synthesis is known as "behavioral" or "functional" simulation. Since only the source code is available and effectively no information about how the design is actually going to be implemented is available, this type of simulation is used for verifying the function of one or more source modules. This is the most frequently run simulation. There are very few reasons why every module of VHDL code shouldn't be verified with this type of simulation.
B. Post-synthesis simulation runs the netlist version of the design. The netlist contains more of the information regarding the silicon resources (i.e. the delays through each silicon resource), but still has no idea about their placement; therefore, no information about the routing (read "delays between silicon resources"). With this information, the post-synthesis simulation will differ in timing from the pre-synthesis; however, the behavior should be the same. The major effect here is that now some of the delays

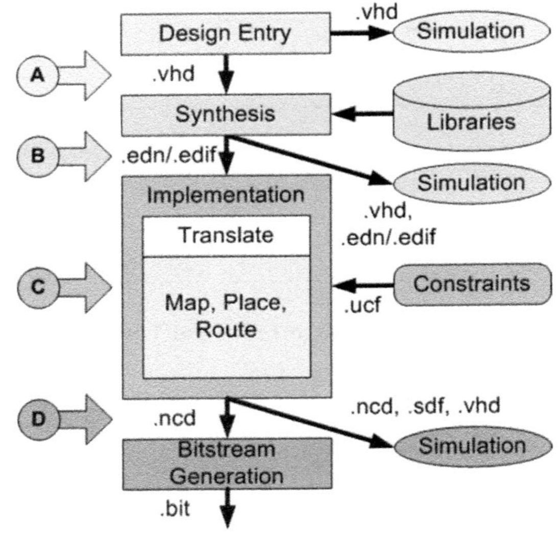

Figure 2.3: Typical Simulation Points During Development.

through the silicon resources are known, and to check the fidelity of the netlist generation.

C. Post-map simulation is rarely run. More knowledge about the specific implementation strategy, and occasionally the physical placement is known and this information is used to generate more detailed, but still approximated, timing simulation for the design.

D. Post-Place-and-Route simulation is the most accurate simulation since all of the delays (both through the ultimate selection of the silicon resources, its location in the FPGA, and the routing among the various resources) are known. This is usually of "sign-off" quality, that is, if the design fully simulates, passes timing analysis,[5] and the test bench is valid, then the FPGA should function as expected.[6] Post-Place-and-Route simulation is the least frequently run simulation as it is the most time-consuming – both in processor time and in writing a quality test bench.

As a general rule, design performance should NOT be determined using simulation. A much better tool is provided in the form of a static timing analysis tool.

Most engineers skip the Post-Place-and-Route simulation, instead relying on functional simulation, static timing analysis, and bench testing.

---

5 Timing analysis is typically performed during implementation phase and confirms that the defined performance constraints are met.

6 For the record, a *good* test bench is difficult and usually quite time-consuming to write which is why many engineers skip this important step and choose to take their design directly to the bench. There are, of course, a number of instances where it is not feasible to write a complete test bench and this will be discussed later.

2.4 Implementation

Implementation is the process of converting one or more netlists into an FPGA-specific pattern. This process is broken down into three basic substeps: translation, mapping, and place-and-route. Refer back to Figure 2.1 (Typical Xilinx FPGA Development Flow), which shows the place and route as separate tasks as it is important to understand the difference between placement and routing.

2.4.1 Translate

The job of the translator is to collect all of the netlists into one large netlist and verify that the constraints map to signals. Any missing modules (which are not detected by the synthesis tools) will be flagged at this stage. Translate also compares the netlist with the user-defined constraints and reports any missing signals.

2.4.2 Map

The traditional role of the mapper is to compare the resources specified in the single grand netlist produced by the translate stage against the resources in the targeted FPGA. If there are insufficient or incorrect resources specified, the mapper will display an error and halt implementation.

As performance is often of concern, modern mappers do more than just compare resources, they may have a limited ability to change the type of resource specified in the netlist (without changing its functionality) to better use the silicon resources. Some mappers will also pre-place certain silicon resources thus taking over some of the role of the placer.

2.4.3 Place and Route

Place and Route is an iterative process which first attempts to "place" the silicon resources, then "route" the signals among the resources such that the timing constraints are met. This stage of implementation makes heavy use of the User Constraints File (UCF),[7] which specifies the maximum duration it may take to move a signal from one silicon resource to another.

[7] A discussion of constraints is worthy of another book. Suffice it to say that the bulk of the constraints have to do with time, while other constraints have to do with forcing locations such as which pins will have which signals or how to group logic into a region within the silicon.

2.5 Bitstream Generation

Almost there! The final step is converting the placed and routed design into a format that the FPGA will understand. Bitstreams can be generated so that they can be directly loaded into an FPGA (usually via JTAG[8]), or formatted for parallel or serial PROMs (Programmable Read-Only Memory).

[8] JTAG – Joint Test Action Group. This group created a process known as "Boundary Scan" which is frequently used for testing circuits in situ. For now, consider this term to be synonymous with "programming cable". Xilinx also uses this same cable to perform monitoring and diagnostics via the ChipScope Pro tool.

CHAPTER 3

Loop 1 – Going with the Flow

This first pass presents the "30,000 foot view" of the language – what are the key parts, the concepts of signals and concurrency, and some basic concurrent operations that can be performed on signals.

3.1 The Shape of VHDL

Let's take a look at the typical parts of a single VHDL module. A module contains two basic parts – one covers the connections made between this module and other modules (the "outside world" which may be the pins on an FPGA package or other modules within the design) and the other part, an unambiguous description of the behavior of the module. Modules may be combined in parallel or hierarchically (see Figure 3.1).

This is a good place to briefly discuss information "scope". Scope is the location or realm that certain information can be accessed from. In the case of a module, the only information from the outside world or higher level that this module can access (either see or modify) must be listed in the *entity* statement. Other information can be defined in both the entity and the architecture sections and can be used freely within this module. If modules hierarchically lower than this module are to share this information, it must be explicitly declared in that module's entity statement.

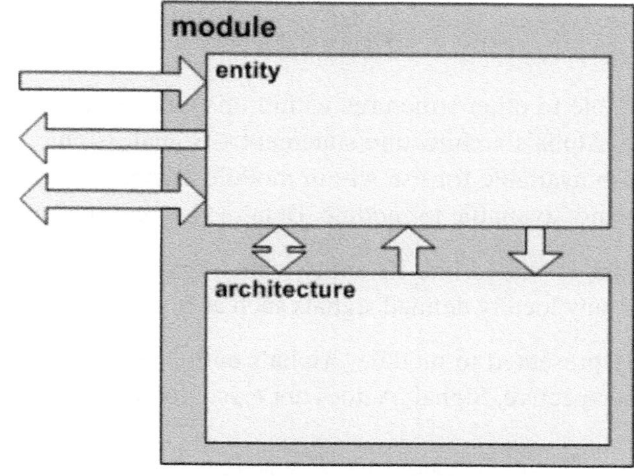

Figure 3.1: The Shape of a VHDL Module.

Figure 3.2 shows a VHDL module comprised of an entity statement and an architecture statement. Only information explicitly listed in the entity is available to the architecture. Within the architecture other information may be defined and used within the architecture and shared with lower-level modules. In order to share this information with lower-level modules, the information must be explicitly listed in the "instantiation[1]" of the lower-level module.

For those readers with a software background, think of lower-level modules as subroutines. The "instantiation" is equivalent to the subroutine "call" and the signals listed in the "instantiation" are equivalent to the subroutine's argument list.

Scoping example – given the following diagram, what signals can be seen in the architecture in module_Alpha? Module_Beta?

Module_Alpha: notice that Signal_A is passed into module_Alpha's entity, which makes this signal "available" for use within module_Alpha. Since it was defined in the "calling"

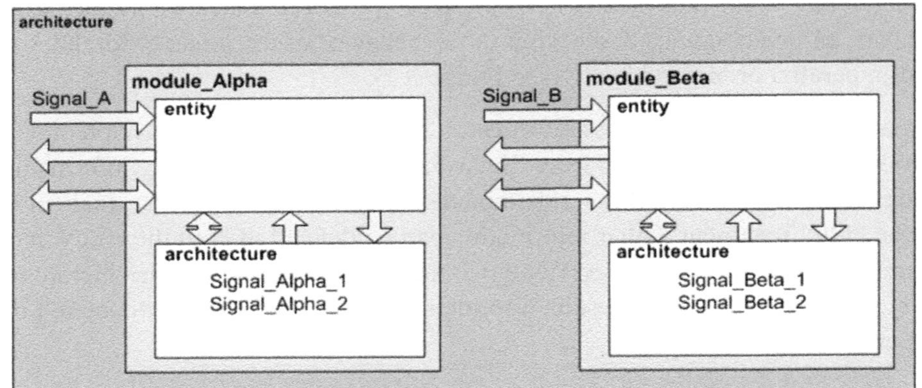

Figure 3.2

architecture, it is available to other structures within this architecture. Two other signals are defined inside module_Alpha's architecture statement – Signal_Alpha_1 and Signal_Alpha_2. These are also available for use within module_Alpha, but only within module_Alpha and are not available to module_Beta or the "calling" module.

Module_Beta has Signal_B passed into its entity, which make it available to module_Beta's architecture along with any locally defined signals such as Signal_Beta_1 and Signal_Beta_2.

Since Signal_A is only presented to module_Alpha's entity, but not to module_Beta's, then, from module_Beta's perspective, Signal_A does not exist. So, to wrap up the basic "shape" of

[1] Instantiation is the explicit reference to another module. A thorough discussion of instantiation and inference is provided later.

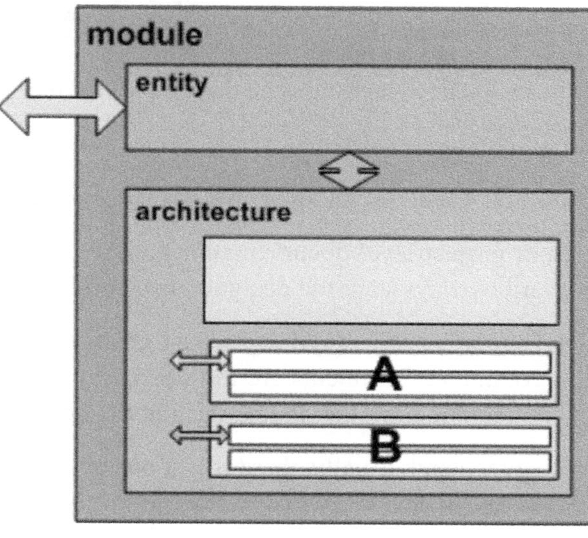

Figure 3.3: Hierarchical Structuring with Entities and Architectures.

VHDL, we can say that a VHDL design is comprised of a single design module which may instantiate multiple modules within it. Each module has the capability of instantiating lower level modules thus enabling a full hierarchical design as the example in Figure 3.3 illustrates.

3.1.1 *The Many Levels of Comments*

Why begin the discussion of VHDL syntax with commenting?

Commenting is a grossly underrated aspect of any design! Without good (read: excellent) documentation, projects flounder. The more sophisticated the project, the greater the need for good documentation. Project-level documentation (specifications) serves as a target to which all code must make a straight line – it provides focus and prevents "feature creep[2]" and guides the development of the project. Code comments and following good design practices help by limiting the scope of what a piece of code must do. Often designers will code in such a fashion that obscures the meaning behind the code making it nearly impossible to maintain – even for the author of the code.

[2] Feature Creep occurs because designers are clever – they think to themselves, if I do it this way then I can expand the capabilities of this design! This has the nasty impact of causing project delays, enlarging the size of the design, and introduces other problems. This is a far more prevalent problem in software than in hardware; however, it is still something that needs to be guarded against.

> **Maxim**: Clarity of thought brings Clarity of code.
> **corollary**: If you can't explain something clearly and concisely, you shouldn't code it.
> **corollary**: A module should be a single thought, that is, it should be described in a single, short sentence.

Maxim 3.1: Clarity of Thought Brings Clarity of Code.

Let's leave the discussion of project-level documentation for the topic of another book and start with discussing the various ways you, the designer, can comment your code.

First, comments in VHDL are really easy – anything to the right of a double dash "--" until the end of a line represents a comment. Comments are terminated by the end of the line which implies that multi-line comments require the double dash at the beginning of each line.

VHDL is a file-oriented language and, usually, there is just one entity and one architecture[3] described within a file. Here is the first place a comment can go – right at the very top of each file. This file level comment should include, at a minimum, the following information:

1. The project name
2. The designer(s) name(s)
3. A version number[4]
4. A history of revisions
5. A status of the state of the code (in development, preliminary, verifying, final etc.)
6. A description of what just this module does including references to any algorithms used, a textual description of the process, etc.
7. A description of the incoming and outgoing signals including what they do and the type of signal they are (i.e., integers, std_logic_vectors, etc.). If this is the interface-level module (i.e., connects to the outside world) then any special information as to pin placement, voltage levels, etc., is desirable. Alternatively, you can list the UCF which contains the pin placement constraints.
8. A description of any unique condition which may require a special constraint or tool setting to implement properly
9. A listing of dependencies – what does this module "call"
10. Anything else you feel important to communicate to someone reading your source code (which will probably be you in 6–9 months and I can almost guarantee that you won't remember a darn thing about the details of this module!)

Single line comments – any time that there are a group of similar statements performing a series of operations to generate a single, specific result, a single line comment (or multiple lines) can

---

[3] VHDL allows for multiple architectures per entity, this is an advanced topic and left for another treatise.

[4] Some Revision/Source Control tools manage versioning. This may be omitted if you are managing versioning outside the code.

be very helpful for describing what a bunch of lines of source code do. For example, if there are a collection of signal definitions, then a useful line of comment might be "-- the next signals are used to move information from module A to module B". This helps things stay organized.

In-line comments – sometimes VHDL (or the algorithm you are using) can be cryptic. A comment hanging on the end of a line might be useful to explain WHY this line was coded the way it was. Unless there is some very complex and convoluted coding going on a comment such as "this line of code increments a counter by one" is next to useless. A more valuable comment might be "this signal is incremented to adjust for something else that is happening elsewhere in the circuit" (but please be more specific than this!).

```
--
-- example of a bad commenting style
-- make a count enable
dayCountEnable <= '0', '1' when (overflow = '1'); -- hold dayCountEnable at
                                          logic low unless overflow is high

--
-- example of better commenting style
-- generate a count enable for the Count Days module
-- this event occurs when count seconds per day module overflows
dayCountEnable <= '0', '1' when (overflow = '1'); -- enable the day counter to
       -- increments when all the seconds in a day have been consumed
```

Snippet 3.1: Example of Commenting.

The first example of bad commenting includes a useless comment before the statement of "make a count enable". Isn't it clear that it's a count enable from the signal name? In the spirit of bad commenting, the in-line comment "hold dayCountEnable at logic low unless overflow is high" describes WHAT the previous statement does, but not WHY. Although this comment might seem useful at the moment while you are learning VHDL, the syntax will become quite familiar and this comment will simply be redundant.

The second commenting example indicates WHY the following line of code exists and when it is expected to occur. The in-line comment says about the same thing and could be omitted with no information lost.

The last type of comment we'll look at is the variable name itself. Some variable or signal names are slightly obvious just because they are used in a specific context so often, such as *i*, *j*, or *k* for a loop index or *x*, *y*, *z* for Cartesian co-ordinate representations. Variables named x12UQ are next to meaningless unless there is a thorough data dictionary. Even then, the reader will have to flip back and forth to understand that this signal might mean "the 12th tap for the Upper Quadrant filter". Since VHDL typically supports 128 letters of significance (VHDL is case insensitive so *position* is the same as *Position* is the same as *POSITION*), try a name such as upperQuadrantTap12 (which uses a Java style naming convention) or upper_quadrant_tap_12 (which is more of a C/C++ naming convention and more commonly used in industry).

The C/C++ naming conventions are typically used with VHDL as the mix of upper and lowercases is generally lost as VHDL converts everything to lowercase during processing. This makes the names difficult to read when using tools that rely on the netlist.

Best Design Practices 3.1 – Typical Naming Conventions for Signals and Variables

1. All uppercase - constants
2. First letter uppercase, everything else lowercase - rarely used
3. All lowercase - signals, variables, label names, functions, processes, procedures, and keywords (context coloring from the text editor provides distinction between keywords and other types)
4. One of many signal-naming conventions: <group association>_<specific signal function>_<clock domain>

Where <group association> represents a module or concept that the specific signal function is a member of.

Another advantage to naming signals this way is that many tools list signals by name and by using a common group association at the front of the signal name, signals that belong together are grouped alphabetically.

<specific_signal_function> represents what the signal is used for; <clock domain> – when more than one clock domain is used in a design, it is often helpful to append the name of the clock domain to the signal so that it becomes easy to see where clock domains are crossed and which signals are related to each other.

This final "field" is useful when your design contains more than one clock domain. This helps to locate clock domain crossings which must be handled carefully in order to avoid meta-stability problems.

As an example, this provides a distinction between the empty flag of one First-In First-Out (FIFO) against another:

INBOUND_FIFO_empty_clkDataIn vs. OUTBOUND_FIFO_empty_clkDataOut

This example also shows that the INBOUND_FIFO_empty is related to the arriving data clock which may be different from the outbound data clock of which the OUTBOUND_FIFO_empty signal is a member.

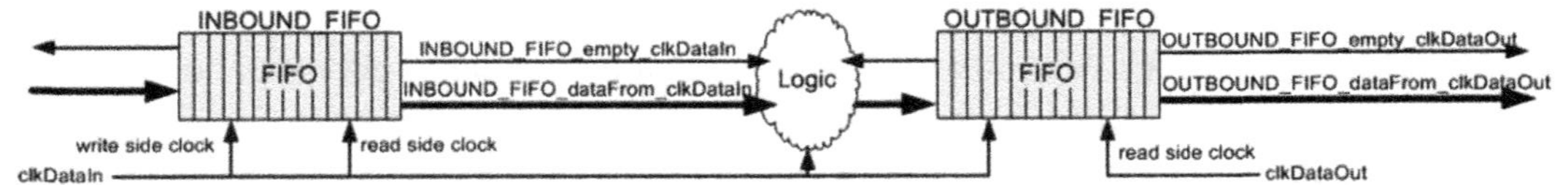

Figure 3.4

From the above example, we can see that the INBOUND_FIFO uses the same clock – clkDataIn for both the read and write sides (that is, for both moving data into the FIFO and pulling data out of the FIFO). This defines this FIFO as a "synchronous" FIFO. Notice that the "empty" status flag and the data leaving the INBOUND_FIFO are both appended with the name of the clock.

These signals arrive at the "logic" (representing whatever this module is supposed to do), including the clkDataIn – making all of the logic part of the clkDataIn domain. As results are generated, they are pushed into the OUTBOUND_FIFO which is still part of the clkDataIn domain.

The OUTBOUND_FIFO differs from the INBOUND_FIFO in that the clock used to add data to the FIFO (clkDataIn) is different than the clock used to extract data from the OUTBOUND_FIFO (clkDataOut). This implies that the OUTBOUND_FIFO is "asynchronous". Notice, again, that the signals coming out of the OUTBOUND_FIFO have clkDataOut appended to them indicating that these signals belong to a different clock domain.

Although the exhaustive list of techniques for how to safely cross time domains is out of the scope of this book, let us just say that without safe time domain crossing circuits, data can become unreliable due to meta-stability.

Best Design Practices 3.2 – Tip
Always cross clock/time domains with a reliable circuit intended for that purpose!

Many designers start coding a module, with only pseudo-code (all comments), then backfill with the VHDL that corresponds to the pseudo-code. This makes for an easy way to self-document. This technique can also be extended to any kind of block structure – for example, always write your "if/then/else/endif" statements all at once, then backfill the true and false clauses – this will save huge amounts of time trying to find that missing "else" clause!

Maxim: Comment first, then backfill with VHDL statements.
Corollary: When coding with block structure statements, code the outsides of the block, then backfill.
Corollary: always comment or use a label at the "end" of the block so that you know what it terminates.

Maxim 3.2: Comment First, then Backfill.

3.1.2 Library and Package Inclusion

As we'll see in the third loop, it's good to reuse code! Libraries and packages are two common mechanisms for accomplishing code reuse. Both Libraries and Packages may be created by the user.

Libraries are collections of one or more packages and may be generated by the user, produced by the Institute of Electrical and Electronics Engineers (IEEE) as part of the VHDL standard (or a proposed standard), or provided by a third-party vendor.

Packages typically contain component definitions, constants, functions, procedures, and user-defined data types, as well as other items. Commonly used items will be identified later in this Loop 3.

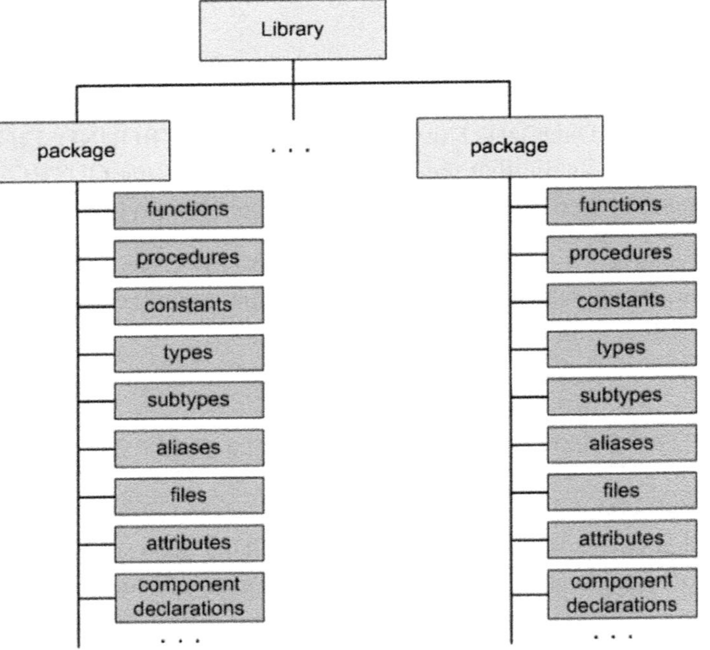

Figure 3.5: Hierarchy of Libraries and Packages.

Libraries and Packages are introduced here for two reasons. First, when used, they are the first statements in the file, and second, the most commonly used data type for FPGA development is the std_logic type, included in the IEEE libraries. Libraries and Packages will be described in more detail in Loop 3, but for now, treat this as a mechanical step and include the first three lines shown in Snippet 3.2 in each module that you write.

Using libraries is an easy two-step process. First, the library itself is declared. This will tell the synthesis tool that this specific library needs to be located and accessed. The exceptions to this are the "Work" library that represents the location to which all compiled code is located[5] and the "STANDARD" library which contains basic data types. Both libraries are available by

[5] It is possible to specify other locations for the code to be compiled to; however, this is not frequently required and is left to the reader to continue his/her study of the specific synthesis tool vendors' user manual.

default. Next, a "use" statement is specified. This indicates which package within the library and the elements within the package should be loaded.

```
1: library IEEE;
2: use IEEE.STD_LOGIC_1164.ALL;
3: use IEEE.NUMERIC_STD.ALL;
4:
5: library UNISIM;
6: use UNISIM.VComponents.FDC;
```

Snippet 3.2: Example Usage of Libraries.

This example shows the use of two libraries – the IEEE library which is typically provided along with the language environment and is constant among implementations. The other library, UNISIM, is provided by Xilinx, an FPGA vendor, and contains the component definitions for the FPGA's primitives.

The "use" statements tell the synthesis tools which package and portion of the package to make available during synthesis.

In the example above, "all" is specified for both of the IEEE packages (lines 2 and 3) and makes everything with the package named STD_LOGIC_1164 and NUMERIC_STD accessible to this module. Line 6 illustrates how only a single component may be made available by calling out its name instead of specifying "all".

As a general rule, standard libraries are accessed with the "all" statement. Calling out specific items within a package is typically used only when there are conflicts between packages. If this occurs, designers will usually create aliases which describe the full path through the library and package to the component that is needed thus avoiding a conflict. This is a bit more complex a topic and is deferred until the discussion of aliases.

3.1.3 Entity

The entity is a "primary" structure, that is, it is required in order for the architecture (behavioral description) to exist. The entity acts as the interface between this module and any other modules or the outside world. It contains the name of the module and a list of signals with a description of what those signals look like.

Snippet 3.3 shows an example of an "Entity" statement with the name "exampleForm". It connects to the outside world (either another module or the pins on the FPGA) with two signals, both of type STD_LOGIC. We'll cover STD_LOGIC and other data types shortly, for now just consider it a binary signal (bit). Signal "*a*" is arriving into the module and the module must drive signal "*b*".

```
entity exampleForm is
  Port (a : in STD_LOGIC;
        b : out STD_LOGIC);
end exampleForm;
```

Snippet 3.3: Example of a Simple Entity Statement.

Please bear in mind that this is an example of a SIMPLE entity construct, as we continue our iterations of this material we will see more capable versions of the entity structure.

Every module must have one and only one entity.

3.1.4 Architecture

The architecture is a "secondary" or optional structure. It may seem strange to have an entity (a definition of the incoming and outgoing signals) without having a behavior associated with it – and certainly final production code is not likely to have this structure, however, during structured code design it may be desirable to have only the entity during the "roughing-in" stage of coding where the architecture(s) may be added later.

The architecture is where the description of the circuit occurs. This is further broken down into two regions. The "what is to be used" (definitions) area resides between the "architecture" statement and the "begin" statement. This portion defines the various signals, constants, functions, procedures, and other modules used in this module.

All of the signals listed in the port statement of the entity for the architecture are available for use by (within) that architecture and, for the purposes of discussion, can be treated as special case members of the "what is to be used" area. The "special case" means that inputs into the module can only be read (consumed) and cannot be written to. Signals that are designated as output from the module can only be driven, they cannot be used in conditionals or to generate other values.

The "how it is to be used" region which is between the "begin" and "end" describes how these signals and modules are connected together and their behavior.

```
architecture Behavioral of exampleForm is
     begin
     end Behavioral;
```

Snippet 3.4: Example Architecture Statement.

Snippet 3.4 shows an example of an empty architecture statement that associates with the Entity snippet shown in Snipper 3.3. These are tied together by the entity name "exampleForm". The word "Behavioral" is the name of the architecture. Is the name "Behavioral" significant? Not really, only in the sense that it should describe what is unique about this particular architecture – especially when multiple architectures are available for an entity. Perhaps an architecture is coded in a purely structural manner, then, the architecture might be named "Structural", or for that matter, "Steve" (just to highlight that the language doesn't care what you name an architecture).

There may be more than one architecture per entity, but this requires a configuration statement to resolve. The way that the configuration statement resolves this is through the name of the architecture. Only one architecture per entity can be "active" at a time. Configuration statements, while indispensable for certain tasks, are rarely used when developing FPGA code. Configuration statements are fairly advanced and detailed explanations may be found in other books.

The VHDL language supports an unlimited number of architectures per entity, but most tools provide a "practical" limit of just a few.

Why even bother with multiple architectures? Shouldn't a module only have a single behavior? Well, yes, but…

1. What if multiple algorithms need to be tested – they all will have the same inputs and outputs and should have the same behavior, but they will synthesize differently? Some may be faster; some may use fewer resources, some different approaches, etc.
2. What if the target FPGA has not yet been selected, then any code that instantiates (explicitly calls out) certain silicon resources may need to reside within its own architecture.

Bottom line – do you need to worry about multiple architectures? Not at this time. Relatively few hardware designers outside of academia use multiple architectures. So, just know that it's there if you need it. We'll focus on single architecture per entity examples and projects in this book.

3.1.5 Configuration Statements

This is, like the architecture, a secondary or "optional" portion. This is a more advanced statement which provides considerable value when implementing a design in multiple FPGA families. The configuration statement provides further information to the synthesis tools as to how to resolve naming conflicts when multiple architectures are present. Further discussion of this statement is deferred to a more advanced VHDL book.

3.1.6 Signals, Data Types, Logical Operators

The architecture statement must contain two types of information in order to produce a meaningful design – what signals (data) must exist and how these signals are combined to describe the behavior of a circuit. This is done by introducing statements that do something with information (which eventually get implemented in CLBs[6] or silicon resources) and connecting them together (which turns into routing).

Let's talk first about the data to be worked on…

Signals

Signals are the primary method for moving information within an FPGA. Simply put, signals become "wires" when synthesized. Signals can be of any legal data type and are valid only within the architecture in which it is declared. So, if you have two VHDL files and both define a signal named "data_arrived" within the architecture, then both versions of data_arrived are completely independent – information is not shared between modules except through access via the entity statements and connection in the "calling" module.

```
1: entity module_A is
2:   port (clk : in  std_logic;
3:            . . .
4:          );
5: end entity module_A;
6:
7: architecure BEH of module_A is
8:      signal data_arrived : std_logic;
9:   begin
10:     data_arrived <= '1';
11:       . . .
12:   end architecture BEH;
```

```
1: entity module_B is
2: port (clk : in  std_logic;
3:            . . .
4:          );
5: end entity module_B;
6:
7: architecure BEH of module_B is
8:      signal data_arrived : std_logic;
9:   begin
10:     data_arrived <= '0';
11:     . . .
12:   end architecture BEH;
```

Snippet 3.5

Assuming that both Module_A and Module_B have been instantiated into a design, what is the value of data_arrived? Notice the signal declaration for data_arrived on line 8 for both modules. This signal declaration indicates that a signal with the name of *data_arrived* will exist in both modules, but are completely independent. *Data_arrived* will have the value of '1' in module_A and will have the value of '0' in module_B. No conflict! (Although it may be somewhat confusing.)

Compare this with the same two modules, but having the signal named data_arrived in the port clause of the entity statement…

[6] OK, there are rare examples of where some designs can exclusively use silicon resources other than CLBs, but let's not get too nit-picky – this is a beginner's book!

```
entity module_A is
  port (clk          : in  std_logic;
        data_arrived: out std_logic;
        . . .
        );
end entity module_A;

architecure BEH of module_A is
  . . .
  begin
    data_arrived <= '1';
    . . .
  end architecture BEH;
```

```
entity module_B is
  port (clk          : in  std_logic;
        data_arrived: out std_logic;
        . . .
        );
end entity module_B;

architecure BEH of module_B is
  . . .
  begin
    data_arrived <= '0';
    . . .
  end architecture BEH;
```

Snippet 3.6

How have things changed? The signal *data_arrived* is now in the port clause of the entity statement which makes these signals accessible to the outside world. Even though the signal *data_arrived* is now available outside the module, these are still independent signals and may be used for other purposes. There would be a problem, though, if the *data_arrived* signals from both modules were connected together. Physically, two drivers are connecting to one another, and, in this case, they are driving to different levels. If this could be implemented in an FPGA, there would be a short-lived battle of the output transistors leading to a lot of heat being produced, then damage to the FPGA. Fortunately, the synthesis tools will quickly identify this problem long before a bitstream could be downloaded to the FPGA. This type of error is called "multi-source".

The Naming of Names

Signal naming is straightforward – the first character of the signal name must be an alpha character followed by any number of letters or numbers or the underscore ('_') character. VHDL is NOT case sensitive, that is, to the synthesis tools the signal named *My_First_Signal_Name* is the same as *MY_FIRST_SIGNAL_NAME* and *my_first_signal_name* (see Figure 3.6). So, the naming rules in brief…

- Alpha-numeric characters (A to Z, a to z, 0 to 9) and the underscore (_) character may be used
- First character must be a letter
- Last character cannot be an underscore
- No consecutive underscores
- VHDL is case *insensitive*
- An identifier can be of any length
  - Synthesis tool typically limit this to 128 characters. For the moment, ignore the various data types and initializations. These concepts will be covered shortly. Pay close attention to the naming of the signals.

```
01: -- legal signal names:
02: signal input_from_laser_gyro: std_logic_vector(15 downto 0) := (others=>'0');
03: signal computation_complete : std_logic := 'U';
04: signal numberOfIterationsRequired : integer := 0;
05: signal channel_4_offset_delay : time := 3 us;
06:
07: -- illegal signal names:
08: signal value_of_2**10        : integer range 0 to 1024 := 1024;
09: signal 5channelSummation     : std_logic_vector(7 downto 0);
10: signal in                    : std_logic := 'X';
11: single hyperConnect_         : std_logic := 'Z';
```

Figure 3.6: Examples of legal and illegal signal names.

Line 2: legal. Notice the use of the aggregate (others => '0') to initialize this signal on power-up.

Line 3: legal. Notice the initialization of this signal to undefined. This is typically done to check that the explicit use of reset properly initializes the signal rather than relying on the power-up condition.

Line 4: legal. Notice the initialization of the unranged integer to 0.

Line 5: legal. Notice the initialization of the physical type "time" and its units.

Line 8: illegal due to use of symbols other than alpha/numeric or '_'.

Line 9: illegal since it begins with a number.

Line 10: illegal because "in" is a keyword.

Line 11: illegal due to the signal name ending with an underscore character.

As VHDL is a strongly typed language, every signal must be declared prior to its use. This declaration is placed between the architecture statement and the begin statement of that architecture.

There are a several parts to defining signals.

- The keyword "SIGNAL" (capitalization not important) which indicates what this line of code does;
- A comma separated list of signal names (it is legal to have a single signal name);
- A colon which separates the list of signal names from the signal type;
- The type_name which may either be:
  - An intrinsic type name,
  - A type_name defined in one of the libraries, or
  - A user-defined type name.
- An optional initialization which includes the required symbol ":=" and a value of the proper type (We'll talk more about this as we move into the simulation realm.);
- A closing semi-colon to separate this line of code from the next.

Best Design Practices 3.5 – Tips

1. When naming a signal, use _n to show that a signal or variable is active low.
2. Make every effort to code resets within the FPGA as active high since this generates the most efficient FPGA code.
    a. *Exception*: some resources within the FPGA require active low resets and/or enables. In this case, invert the reset or enable local (i.e. within the same module) as the special resource.

```
SIGNAL signal_name_list : type_name [:= value];
```

Spec 3.1: Signal Declaration.

The second part is that of the signal assignment. As with most programming languages, the target of the assignment resides on the left-hand side of the signal assignment association ("<=") symbol and the generation of the signal value resides on the right. While this appears to be a bit awkward for an assignment, it serves to differentiate a signal assignment from a variable assignment (":=") and from a conditional test for equality ("="). For one signal to be assigned to another (or one equation to be assigned to a signal) a very simple, but important rule must be followed: whatever is on the right side of the assignment operator must be the same size and same type as that on the left side of the operator.

```
1: architecture Behavioral of exampleForm is
2: signalc : std_logic;
3: begin
4:  c <= a;
5:  b <= c;
6: end Behavioral;
```

Snippet 3.7: Example of a Signal as a Concurrent Statement.

Snippet 3.7 shows that the signal named *c* is defined as a std_logic data type (which we'll describe shortly) on line 2. Line 4 shows how the signal named *a*, which we can assume was defined in the port statement in the entity, is assigned to our new signal *c*. Finally, line 5 shows how the internally defined signal *c* can be assigned to drive an output signal *b* (also assumed to have been defined in the port statement of the entity).

Signal assignment can take many forms from the simple assignment shown in Snippet 3.7, to logical or mathematical computations, to conditional assignments, and to a non-synthesizable construct used in developing test benches which include information as to when to make the assignment and a number of other aspects. We will cover many of these assignments as we revisit this topic later in this book.

Data types

A data type is the way information is represented in the code. This distinction is made because once the design hits the FPGA, it's all ones-and-zeros. As with any digital device, data is represented intrinsically as binary – logic true/high or logic false/low. By gathering these individual bits together and treating them in a certain fashion, they can represent other types of information such as integers, floating-point numbers, and characters.

Formally, a data type is comprised of two parts: that of the values that the data type can take on and by a set of operations that can be performed on that data type.

The official VHDL "language" (without libraries) offers only a few data types – most VHDL designers rely on a "package" or collection of extensions which provides the std_logic/ std_logic_vector type.

Scalar types

The first kind of data types is called "Scalar types". This means that they are single-dimensional values such as an integer or real number. These scalar types are broken down into four basic groups: enumeration types, integer types, floating types, and physical types.

Enumeration types. Enumeration types are built up from explicitly defined values. This is to say that there is no natural range such as integers which have a fixed pattern between any two values.

As a simple example, let's look at the intrinsic (built into the language) VHDL type of the bit. A "bit", not surprisingly, is a logic '1' or a logic '0'. Since the "bit" can only take on one of two values, a definition of an enumerated type for bit looks like the one in Figure 3.7.

```
type BIT is { '0', '1' };
```

Figure 3.7: Example of Enumerated Type for bit.

So, a bit may take on one of these values or the other. Any other assignment is invalid and will be flagged by the compiler as an error.

Certainly most synthesizable logic can be performed with the bit type; however, we know that it may be possible to have a digital signal in high-impedance mode, or, when simulating, not known or don't care about the value of the signal. A more versatile and strongly preferred

```
TYPE std_logic IS ( 'U', -- Uninitialized
            'X', -- Forcing Unknown
            '0', -- Forcing 0
            '1', -- Forcing 1
            'Z', -- High Impedance
            'W', -- Weak Unknown
            'L', -- Weak 0
            'H', -- Weak 1
            '-' -- Don't care
      );
```

Figure 3.8: Example of Enumerated Type for std_logic.

mechanism to represent logic is the "std_logic" data type. This type is defined in the IEEE 1164 standard[7] similar to the one in Figure 3.8.

As we can see from the above example, *std_logic* can take on one of nine different values:

- '1' – the usual indicator for a logic high signal
- '0' – the usual indicator for a logic low signal
- 'H' – weak '1'
- 'L' – weak '0'
- '-' – don't care
- 'U' – undefined – used most frequently in simulation to mark an unintialized or undefined signal level
- 'Z' – high impedance – often used in simulating devices outside of the device under test (FPGA) or tri-state outputs
- 'W' – "weak unknown"
- 'X' – "forcing unknown"

Typically, we only use '1', '0', 'Z', and occasionally 'X' when we write code for synthesis (inclusion into an FPGA) and the other values are more commonly seen in simulation. *Note*: although VHDL is a case insensitive language, when enumerated types are enclosed by single quotes, the case does become important!

As a designer, you will most likely use your own enumerated types for naming various states within a state machine.

The other part of an enumerated type is the operations that can be performed on them. In the case of the std_logic type, the IEEE library 1164 defines all of the operations which can be performed on a std_logic type. These include the usual logic operations (and, nand, or, nor, xor, and not). Other libraries (such as IEEE numeric_std) extend these capabilities.

When you define your own types, you may need to also define the operations to be performed on them (but not always!)

[7] Which means that the library and the associated use clause must be present at the top of each VHDL module that uses the std_logic type.

Here are two more frequently used (and predefined) enumerated types that you will likely require…

Character – one of 256 characters defined in the ISO 8859-1 character set. Characters, like Booleans, are generally used more often in test benches than in synthesizable designs.

Boolean – comprised of the values "true" and "false". Although synthesizable, Booleans more commonly see application in test benches. Typically Booleans are used to temporarily store the results of complex comparisons.

```
type BOOLEAN is { false, true };
```

Spec 3.2

And now for an example:

```
1: signal is_operational : boolean := false;
2: signal power_applied, no_errors, initialized : std_logic;
3: . . .
4: is_operational <= (power_applied='1') AND (no_errors='0') AND
                      (initialized='1');
5: do_something <= '0', '1' when (is_operational);
```

Snippet 3.8

Line 1: between the architecture and its begin statement, *is_operational* is declared as a Boolean and initialized on power-up to *false*

Line 2: between the architecture and its begin statement, other signals are defined as single-value quantities

Line 3: other statements including "begin" which ends the declaration area of the architecture and enters the behavioral description of the architecture

Line 4: the single-value quantities are combined and the result will be Boolean as each test produces a Boolean quantity

Line 5: the Boolean quantity *is_operational* is used to set the value of *do_something*

Integer types. Integers range from at least −2,147,483,647 to +2,147,483,647 (+/− 2^32) inclusive. The standard also states that any given implementation of VHDL may exceed this value, but no legal implementation may define an integer any less than this range.

When used in FPGAs, integers are usually given a subrange. Limiting the range of an integer reduces the amount of silicon resources required to implement that value and is an intelligent way of error checking.

Consider: if subranges were not applied to integers, then every integer would have to be represented by the number of bits used to represent the largest possible integer, which is at least 32 bits. Each register used to hold an integer would be 32 bits, every data path carrying an integer from one point to another would be 32 bits, etc. Not only would this likely force a design into a larger FPGA, but the process of meeting timing would be hampered due to the tools having to route far more signals than necessary.

Consider: if a three-dimensional steering mechanism were being constructed, then azimuth would be limited to a range of 0–359 or −180 to +180 and elevation would be limited to a range of −90 to 90. Not only would this require only 17 bits instead of 64, but simulation could test for values outside these ranges and help the designer detect issues with the design long before field tests were run.

Figure 3.9 is an example of a common signal definition for a byte-sized integer datum…

```
signal byte_value : integer range 0 to 255;
```

Figure 3.9: Signal definition for a byte sized integer.

Floating types. The only predefined floating type is that of the REAL. As with the integer a minimum range is required to be supported (in the case of REAL this range is −1.0E38 to +1.0E38 inclusive), but any particular implementation may legally extend this range.

Floating types are not intrinsically synthesizable, that is, they are limited to simulation only. That said, the IEEE does have a proposed synthesizable floating-point standard which does work, however, the implementation performance is not especially efficient. There are a number of techniques for managing non-integer representations including fixed-point techniques. The details of how this can be done is the topic for a more advanced book.

Physical types. The last scalar type – that of the physical type – is provided to give context to real-world quantities. The only predefined physical type is that of time; however, numerous libraries are available to define other physical types such as distance, voltage, and current.

Users can also define custom physical types.

Similar to floating types, physical types are not synthesizable, but rather intended for simulation. Using physical types, board-level and system-level simulations are possible (including modeling of analog-to-digital converters, transistors, and other devices).

The most commonly used physical type is that of time which is almost always used in test benches to generate clocks and other stimuli.

Table 3.1 Summary of commonly used types.

Type	Values	Package
std_logic	'0', '1', 'Z', 'U', 'H', 'L', 'W', 'X', '-'	std_logic_1164
std_logic_vector	Array of std_logic	std_logic_1164
unsigned	Array of std_logic	numeric_std
signed	Array of std_logic	numeric_std
integer	$-(2^{31}-1)$ to $(2^{31}-1)$	standard
real	−1.0E38 to 1.0E38	standard
time	1 fs to 1 h	standard
Boolean	true, false	standard
character	256 characters as defined in	standard
String	Array of characters	standard

Composite types

Composite types define collections. These collections could be an ordered grouping of the same type of scalar object – arrays, or of different types of scalar objects and/or other composite types – records.

Let's start with arrays:

As was just mentioned, arrays are used for gathering multiple items of the same data type together. For example, gathering 32 std_logic signals together forms a bus.

```
array  <index_constraint> of <element_subtype_indication>
```

Spec 3.3

Typically the <index_constraint> is just the range of the array, such as (1 to 10) or (0 to 99), while the <element_subtype_indication> defines the kind of data that is to be stored in the array.

Let's take a short detour into the <index_constraint>. Typically the index constraint is given as a range. This range is usually from a lower number to a higher number when used with a standard array. For example, if an array of 10 integers is defined to start at index zero, then it would be written as "array(0 to 9) of integer". Indexes can cover any range – including negative indexes: "array(−180 to 180)" is a perfectly legal construct.

Std_logic_vectors are defined to be arrays of std_logic; however, the convention is to specify the higher index to the right which is useful when converting a std_logic_vector to an unsigned/signed type or integer. A commonly used array of std_logic is to define a byte: "array(7 downto 0) of std_logic". Of course the std_logic_vector type is a shorthand for this and is equivalent to "std_logic_vector(7 downto 0)".

Arrays can't be applied directly to a signal, they must first be defined as a type or subtype, then applied to the signal.

Some examples:

```
1: type intAry100 is array (0 to 99) of integer range 100 to 100;
2: signal randNumbers : intAry100;
3: type slvAry10 is array (10 downto 1) of std_logic_vector(3 downto 0);
4: signal stuff : slvAry10;
```

Snippet 3.9

Line 1: a type is created to reference the array. This particular array contains only integer ranging from −100 to 100 and is 100 integers in length. The lowest index into this array is 0.

Line 2: a signal named *randNumbers* is defined to be of type intAry100 which was defined on the previous line.

Line 3: another array type is defined, this time having a size of 10 elements each being a 4-bit std_logic_vector. Notice the use of "to" and "downto". Convention has it that arrays are generally designated from low index "to" high index, although VHDL does not put such restrictions on the designer.

Line 4: the array *slvAry10* becomes the type for signal *stuff.*

Notice that neither of the arrays above is initialized. Therefore, if this code is synthesized, then on power-up they will take on whatever the default initialization values are. Best Design Practices calls out initializing the arrays which is done as follows:

```
1: type intAry100 is array (0 to 99) of integer range -100 to 100;
2: signal randNumbers : intAry100 := (others=>-100);
3: type slvAry10 is array (10 downto 1) of std_logic_vector(3 downto 0);
4: signal stuff : slvAry10 (10=>X"F",5=>X"3", 1=>X"A", others=>X"0");
```

Snippet 3.10: Initializing the arrays.

Line 2: This line now initializes every member of randNumbers to the value of −100. This type of assignment is known as an "aggregate" and will be covered in more depth shortly.

Line 4: another example of an aggregate used to initialize the signal *stuff.* In this case, the aggregate specified that index 10 receives the value hexadecimal 'F', index 5 receives the hexadecimal value of '3', and index 1 receives the hexadecimal value of 'A'. All other members of the array are initialized to 0.

When arrays are used they can be accessed in one of three ways:

1. In their entirety (the whole array is used in a single assignment or test)
2. A single element of the array is used
3. A subrange of the array is used.

```
1: type circleArray is array (0 to 360) of integer range 0 to 10;
2: signal circle1 : circleArray;
3: signal circle2 : circleArray;
4: signal target  : integer range 0 to 10;
5: ...
6: circle2 <= circle1;
7: target  <= circle1(27);
8: circle1(85 to 95) <= circle2 (0 to 10);
```

Snippet 3.11

Line 1: circleArray is created as a user-defined type of an array ranging from 0 to 359 where every element in the array is an integer having a value of between 0 and 10.

Lines 2,3: circle1 and circle2 are defined as circle arrays.

Line 4: target is defined as an integer between 0 and 10.

Line 6: circle1 assigned to circle2 in its entirety. Note that no indexes are used.

Line 7: *target*, which is a range-limited integer gets a single value from array *circle1*. This assignment is possible because each element in the circleArray is defined in the same fashion as *target* is.

Line 8: 11 elements from circle2 (index 0 through index 10) are copied into circle1 beginning at index 85. This is a legal assignment because the number and type of items are the same.

As we mentioned in our brief discussion of enumerated types, the most frequently used type for representing signals within an FPGA is the std_logic type. The std_logic_1164 package predefines a structure which may contain one or more of these std_logics. This grouping is known as a std_logic_vector. Std_logic_vectors are usually designated with a "downto" so that the least significant bit is in the rightmost position. While std_logic_vectors have no specific positional meaning, they can take on meaning when converted to other types.

This means that just because a bunch of signals is lumped together, doesn't give it an intrinsic meaning! A std_logic_vector of "1001" is NOT equal to 9 – it is only an un-interpreted string of ones-and-zeros. We'll cover how to have std_logic_vectors interpreted as integers shortly.

Slices and aggregates

Since we are discussing arrays, we must also discuss how to assign values to and use pieces of arrays. The slice is method of specifying a contiguous region within an array, and an aggregate is a mechanism to assign or access non-contiguous values within an array.

Slices are easier to understand, so we'll begin with an example:

```
 1:  type short_array_of_integers is array (0 to 7) of integer;
 2:  type really_short_array_of_integers is array (0 to 3) of integer;
 3:  type memory_structure is array (0 to 255) of
        std_logic_vector(7 downto 0);
 4:  constant fibonicci : short_array_of_integers :=
       (1, 1, 2, 3, 5, 8, 13, 21);
 5:  constant primes    : short_array_of_integers :=
       (1, 2, 3, 5, 7, 11, 13, 17);
 6:  signal     memory : memory_structure := (others=>(others=>'0'));
 7:  signal     a      : integer := 0;
 8:  signal     b      : short_array_of_integers := (others=>0);
 9:  signal     c, cc  : std_logic_vector(7 downto 0) := (others=>'U');
10:  signal     d      : really_short_array_of_integers := (others=>0);
11:   . . .
12:   a <= fibonicci(4);
13:   b <= primes;
14:   c <= memory(127);
15:   d <= really_short_array_of_integers(fibonicci(2 to 5));
16:   cc(1 downto 0) <= "11";
```

Snippet 3.12

Code analysis:

Line 1: a short_array_of_integers which is an array of integers having eight members with the first member at position 0 is created.

Line 2: a similar array that has only four integer members.

Line 3: another array which defines a memory_structure of bytes (built of type std_logic_vector) having 256 members with the first member indexed to position 0. Convention has it that std_logic_vectors are defined using the "downto" notation. The rationale behind this is so that the "least significant bit" is in the rightmost position. Remember that std_logic_vectors, like any array, are collections of signals and have no positional meaning associated with them. Only conversion and casting functions like "to_unsigned" give the bit positions meaning.

Line 4: a constant called *fibonicci* is defined as a short_array_of_integers and is initialized. Notice that each array element has a corresponding value, that is, there are eight members of the initialization list, one for each member of the array. The length of the initialization list must exactly match the number of elements in the array.

Line 5: another constant, this one named *primes* is initialized.

Line 6: signal *mem* is defined and initialized to all zeros. Notice the (others => (others => '0')) initialization. This translates to "for every element in the array assign every element of std_logic_vector that the array is made of to a zero".

Lines 7–10: definition of signals *a*, *b*, *c*, *d*.

Line 12: signal *a*, which is defined as an integer, is assigned the fifth (since the array starts at 0) element of the *fibonicci* array (or value 5).

Line 13: signal *b*, which is defined as a short_array_of_integers, is assigned all the values of *primes*. This is legal since both *b* and *primes* are defined as *short_array_of_integers*.

Line 14: signal *c* receives the last value in the memory array. Since each member of *memory* is an 8-bit std_logic_vector, signal *c* must be of the same type.

Line 15: even though *fibonicci(2 to 5)* defines an integer array of length 4, it cannot be assigned directly to signal *d* as signal *d* is defined as type *really_short_array_of_integers*. Since the length and type is correct, this slice of constant *fibonicci* can be cast as a *really_short_array_of_integer* then assigned to signal *d*.

Line 16: slices don't always have to appear on the left side of the assignment symbol. This line of code sets the 1 and 0 positions in the std_logic_vector *cc* to ones. The other positions in *cc* remain unchanged.

Next we'll contend with the only slightly more complex "aggregate". As the name suggests, an aggregate is a collection of similar items brought together. The key difference between the aggregate and the slice is that the slice can refer to only an individual item or a contiguous region within an array. Aggregates can specify non-contiguous regions in arrays as well as be applied to records (next section; see Figure 3.10).

31	30	29	28	27	26	25	24	23	22	21	20	19	18	17	16	15	14	13	12	11	10	9	8	7	6	5	4	3	2	1	0
1	-	-	-	-	-	-	-	0	0	0	0	0	0	0	0	-	-	-	-	-	-	-	-	-	-	-	-	-	-	H	H

Figure 3.10: Result of aggregate assignment shown on Line 3.

```
1: signal sig :  std_logic_vector(31 downto 0)  := (others=>'U');
2: . . .
3: sig <= (31=>'1',23 downto 16 =>'0', 1|0 => 'H', others=>'-');
```

Snippet 3.13: Example of Aggregate Use (Left-Hand Side).

Code analysis:

Line 1: signal *sig* is defined as a 32-bit std_logic_vector and initialized using an aggregate to push all values not previously defined (which is all of them) to the value of 'U' or undefined – a legal value for std_logic_vector.

Line 3: signal *sig* is assigned using an aggregate. This aggregate reads as follows: place a '1' in position 31, place a '0' in all positions between 16 and 23 inclusive, place an 'H' in positions 0 and 1, and to all positions not otherwise described, fill with a '-' value. See Figure 3.10.

To close out this section, let's look at how aggregates can be the target of assignments.

```
 1: type really_short_array_of_integers is
                          array (0 to 3) of integer;
 2: signal dd    : really_short_array_of_integers := (others=>0);
 3: signal word : std_logic_vector(3 downto 0) := X"C";
 4: signal slW, slX, slY, slZ : std_logic;
 5: signal iA, iB, iC, iD : integer;
 6:     . . .
 7: dd(0) <= 10;
 8: dd(1) <= 21;
 9: dd(2) <= 32;
10: dd(3) <= 43;
11: (slW,slX,slY,slZ) <= word;
12: (iA, iB, iC, iD) <= dd;
```

Snippet 3.14: Examples of Aggregates on the Right Side of an Assignment Statement.

Code analysis:

Lines 1–5: definitions of signals and types used in this example. Notice the use of the character 'i' for the integer signal definitions and the use of "sl" for the std_logic signal definitions. This is referred to as "Hungarian" Notation and is often used in software to identify the type of the variable or signal. This nomenclature can be quite handy to distinguish various forms of signals or variables when conversions are performed.

Lines 7–10: the array dd is filled with specific values.

Line 11: *word* is defined as a std_logic_vector of length 4. Since each member of a std_logic_vector is a std_logic, these individual positions can be applied to any other signal or variable requiring a std_logic type. In this case, to preserve the length, slW, slX, slY, and slZ will be assigned, from left to right, the values in *word.*

Line 12: just as was done in line 11, dd is defined as an array of integers 4 elements long. Signal *iA* is assigned the leftmost integer in the array, in this case 10. Signal *iB* then receives the value of dd(1) which is 21, and so forth.

Records

Let's take a look at the record structure. A record is a collection of different types of scalar objects or other composite types for that matter.[8]

Records are extremely useful for associating or grouping information together, regardless of type.

```
type <record name> is record
   <element declarations>
end record;
```

Spec 3.4

Where the <record_name> is a legal identifier and follows the same naming rules as a signal and <element declarations> is the list of items to be collected into the record.

As with so many other things, an example is worth a thousand words…

Consider a PCIe Transaction Layer Packet (TLP) for a 32-bit address space write:

<table>
<tr><th colspan="8">+0</th><th colspan="8">+1</th><th colspan="8">+2</th><th colspan="8">+3</th><th></th></tr>
<tr><td>7</td><td>6</td><td>5</td><td>4</td><td>3</td><td>2</td><td>1</td><td>0</td><td>7</td><td>6</td><td>5</td><td>4</td><td>3</td><td>2</td><td>1</td><td>0</td><td>7</td><td>6</td><td>5</td><td>4</td><td>3</td><td>2</td><td>1</td><td>0</td><td>7</td><td>6</td><td>5</td><td>4</td><td>3</td><td>2</td><td>1</td><td>0</td><td></td></tr>
<tr><td>R</td><td colspan="2">Fmt</td><td colspan="5">Type</td><td>R</td><td colspan="3">Traffic Class</td><td colspan="4">R</td><td>TD</td><td>EP</td><td colspan="2">Attr</td><td colspan="2">R</td><td colspan="10">Length</td><td>DW 0</td></tr>
<tr><td colspan="16">Requester ID</td><td colspan="8">Tag</td><td colspan="4">Last DW BE</td><td colspan="4">1st DW BE</td><td>DW 1</td></tr>
<tr><td colspan="30">Address[31:2]</td><td colspan="2">R</td><td>DW 2</td></tr>
</table>

Figure 3.11

The specifics of how this header works is inconsequential. What is relevant is that there are numerous fields, each with different types of units. The following record type defines a record for this Memory Access TLP.

Code analysis see Snippet 3.15:

Line 1: FORMAT_TYPES is defined to be one of four enumerated values. From the VHDL designer's standpoint, the order is generally inconsequential. The synthesis tools will assign

[8] Virtually all programming languages have an analogy to this structure – Java call them classes, C/C++ calls them structures, FORTRAN (for you old-timers reading this book) uses common blocks, Pascal calls them records, etc.

```
 1: type FORMAT_TYPES is (noData3DW, noData4DW, data3DW, data4DW);
 2: subtype DWORD     is std_logic_vector(31 downto 0);
 3: type slvArray1024 is array (1 to 1024) of DWORD;
 4: type TLPmemAccess is record
 5:   format         : FORMAT_TYPES;
 6:   tlpType        : std_logic_vector(4 downto 0);
 7:   trafficClass   : integer range 0 to 7;
 8:   digestPresent  : boolean;            -- corresponds to field TD
 9:   poisoned       : boolean;            -- corresponds to field EP
10:   attributes     : std_logic_vector(1 downto 0);
11:   length         : integer range 0 to 1024;
12:   reqID          : std_logic_vector(15 downto 0);
13:   tag            : integer range 0 to 255;
14:   lastBE         : std_logic_vector(3 downto 0);
15:   firstBE        : std_logic_vector(3 downto 0);
16:   address        : std_logic_vector(31 downto 2);
17:   data           : slvArray1024;
18:   digest         : integer;
19: end record TLPmemAccess;
```

Snippet 3.15: TLP Record Example.

binary codes to the enumerated types for synthesis purposes. The binary patterns should not and cannot be used as a comparison against the enumerated types.

Line 2: defines DWORD (for double-word) as a std_logic_vector of 32 bits. This is a subtype rather than a type because std_logic_vector is already a type and we are further restricting the range of the std_logic_vector type.

Line 3: an array of 1024 double-words is defined for future use in carrying the data associated with a TLP.

Line 4: defines the beginning of the record.

Line 5: defines format of type FORMAT_TYPES.

Line 6: tlpType is defined as a 5-bit std_logic_vector.

Line 7: traffic class is typically used as an integer as it may be tested against other traffic classes to decide priority. Remember that integer can be used to test various relationships such as "greater than" and "less than", but std_logic_vectors can only be used to test for equality or inequality. Std_logic_vectors would need to be converted to signed/unsigned or integer type for other relationship testing.

Line 8: digestPresent is a Boolean which indicates if an extra checksum is present at the end of the TLP. Although this could have been implemented as a bit type or as a std_logic type, using a Boolean type provides far more readable (maintainable) code.

Line 9: poisoned is defined as a Boolean for the same reason that digestPresent is.

Line 10: attributes are defined as a 2-bit std_logic_vector.

Line 11: length is a value between 0 and 1023 which will be adjusted later to reflect the range from 1 to 1024.

Line 12: regID is defined as a std_logic_vector since it has several fields within which it is much easier to extract as a std_logic_vector rather than converting it to an integer.

Line 13: tag – an integer between 0 and 255.

Line 14: lastBE – since the individual bits form a mask for the last double-word (32 bits) of the data provided in the TLP, this is implemented as a std_logic_vector.

Line 15: firstBE – same rational as lastBE.

Line 16: address – although address computations may be necessary, which would suggest that an integer might be appropriate, there are two reasons why this is being handled as a std_logic_vector. First, the minimum range of an integer is $-(2^{31} - 1)$ to $+(2^{31} - 1)$ and addresses are generally positive, this would make the upper half of the address range inaccessible unless negative addresses were used and that's just plain confusing! It may be that a particular synthesis tool will support integers which are larger, but that would make the code less portable among synthesis tools. Additionally, PCIe supports 64-bit addressing which is likely to be out of the supported range of any synthesis tool. The second reason is that certain bits within the address may need to be masked, and although this is possible with integer, the code will be more maintainable if std_logic_vector is used.

Line 17 – data – the total amount of data transmitted in any given TLP is defined by the length. Since VHDL does not natively support dynamic memory allocation in synthesis, this array must be set to the maximum possible side of a TLP data field which is 1024 bits. Note that an array is used here since all the members of this field are of the same type – 32-bit std_logic_vectors (DWORDs).

Line 18 – digest – this is an optional field and will exist depending upon the value of digestPresent. It is defined here as an unranged integer. This will cause the tools to use the maximum number of bits available to the synthesis tools, which, at a minimum, is 32 bits. This is a trifle unsafe as it makes an assumption that the synthesis tools will implement only 32 bits. The danger here is that if the Cyclic Redundancy Check (CRC) value which is being calculated internally also uses an unranged integer, then there is a risk that the transmitted CRC (of which only 32 bits are transmitted) may not match the computed CRC (which may be implemented with more than 32 bits). In order to protect against this, the computed CRC should be masked so that only the lower 32 bits are kept.

Line 19: marks the end of the record.

How might we initialize and use such a record? Certainly values can be explicitly assigned one at a time…

```
signal currentPacket : TLPmemAccess;
. . .
currentPacket.format  <= noData3DW;
currentPacket.tlpType <= (others=>'0');
currentPacket.trafficClass  <= 3;
currentPacket.digestPresent <= true;
currentPacket.data(2) <= (31 downto 16=>'0', others=>'1');
```

Snippet 3.16: First Example of Assigning/Initializing a Record.

Code analysis:

Line 1: signal *currentPacket* is defined as a record type of *TLPmemAccess.*

Line 3: the field *format* is identified within *currentPacket* using "dot" notation. Since the *format* field is defined as a user-defined type of *FORMAT_TYPES* it may only receive one of the four defined values for this user-defined type, of which *noData3DW* is one.

Line 4: field *tlpType* is a 5-bit std_logic_vector which is initialized using an aggregate to force all positions within the std_logic_vector to '0'.

Line 5: field *trafficClass* is an integer with a range of 0 to 7 and is assigned an integer value within the specified range.

Line 6: field *digestPresent* is a Boolean and can therefore be assigned *true.*

Line 7: the *data* field in the TLP is an array of 1024 std_logic_vectors of 32 positions each. This line refers to the third element in this array. The aggregate assigns this one element a pattern of 16 '0's in the 16 through 31 positions, and the remaining positions (0 through 15) are assigned a value of '1'.

While this works and is fairly clear, we can use an aggregate to perform the same task:

```
 1: currentPacket <= (format=>noData3DW,
 2:                    tlpType=>"10100",
 3:                    trafficClass=>3,
 4:                    digestPresent=>true,
 5:                    poisoned=>false,
 6:                    attributes=>"10",
 7:                    length=>3,
 8:                    reqID=>(others=>'0'),
 9:                    tag=>45,
10:                    lastBE=>X"F",
11:                    firstBE=>X"F",
12:                    address=>(31 downto 10=>'1',others=>'0'),
13:                    data=>dataArray,
14:                    digest=>16#12345678#
15:                   );
```

Snippet 3.17: Using an Aggregate to Initialize/Assign to a Record.

Code analysis:

Line 12: notice the use of the aggregate to define the std_logic_vector used within the aggregate to assign to *currentPacket.*

Line 14: we're used to identifying hexadecimal values when assigned to std_logic_vectors (X"…"), here is how a hexadecimal integer value is represented. Following the pattern set forward by the hexadecimal integer notation, 2#…# is used to define a binary number, and 8#…# is used to define an octal number.

Note that when using an aggregate, all fields must be assigned unless the "others" clause is used. When the "others" clause is used, the value must be applicable to all the undefined fields. For example, you could NOT use the following attribute to initialize *currentPacket*:

(others => '0') since '0' is valid only for std_logic and has no meaning when applied to integers, Booleans, or user-defined types (see Table 3.1).

Operators

Now that we've covered the basic data types, let's cover what can be done with them.

The first thing to keep in mind is that operators have specific data types that they can work on. Some of these are intrinsic to the language, such as adding two integer numbers, while others are defined in specific packages such as adding an integer to a signed. Not every operator can be used with every data type.

Taking this one step further, not every operation, division for example, is synthesizable.

There are seven categories of predefined operators in VHDL. They are as follows:

- Binary logical operators (and, nand, or, nor, xor, xnor)
- Relational operators (=, /=, <, <=, >, >=)
- Shifting operators (sll, srl, sla, sra, rol, ror)
- Addition operators (+, −, &)
- Unary sign operators (+, −)
- Multiplying operators (*, /, mod, rem)
- Other operators (not, abs, **)

We'll cover the relational operators in the next loop when we introduce comparisons.

Logical operators are the means in which to combine signals (hence the term "combinatorial logic"). These are the basic "AND" gates, "OR" gates, etc.

These operators work pretty much as you'd expect them to – the notation, like most languages, is a little different in places. Where the type "logic" is specified, then it could be of bit/bit_vector type or of std_logic/std_logic_vector type. If std_logic/std_logic_vector is used, be certain to include the STD_LOGIC_1164 package as that includes the definitions for the operators. The result of an std_logic <logical operator> std_logic is going to be an std_logic, while an std_logic_vector can only be manipulated with another std_logic_vector, that is, you cannot "AND" an std_logic_vector with an std_logic. As always, the "same size-same type" rule applies.

"Numeric" in the following context means either an integer or a real number. Where the result is defined as numeric then it is usually real if one of the operands is real, integer if the operands are integers.

---

[9] The shifting and rotating operators are intended for bit_vectors only. The equivalent operation for std_logic_vectors is defined in NUMERIC_STD as shift_right, shift_left, rotate_right, rotate_left. Shift right arithmetic preserves the most significant bit so that the sign is maintained, no equivalent for this exists in the NUMERIC_STD package.

Table 3.2

Operation	Operator	Type of Operand	Type of Result
Exponentiation	**	numeric ** Integer	numeric
Absolute Value	Abs	numeric	Numeric
Compliment	Not	**not** Logic	Logic
multiplication	*	numeric * numeric	numeric
division	/	numeric / numeric	Numeric
modulo	mod	integer **mod** integer	integer
remainder	rem	integer **rem** integer	Integer
unary plus	+	+ numeric	numeric
unary minus	−	− numeric	numeric
addition	+	numeric + numeric	numeric
subtraction	−	numeric − numeric	numeric
concatenation	&	(array or element) **&** (array or element)	Array
shift left[9] logical	sll	logical array **sll** integer	Logical array
shift right logical	srl	logical array **srl** integer	Logical array
shift left arithmetic	sla	logical array **sla** integer	Logical array
shift right arithmetic	sra	logical array **sra** integer	Logical array
rotate left	rol	logical array **rol** integer	Logical array
rotate right	ror	logical array **ror** integer	Logical array
test for equality	=	object = similar object	Boolean
test for inequality	/=	object /= similar object	Boolean
test for less than	<	object **<** similar object	Boolean
test for less than or equal	<=	object **<=** similar object	Boolean
test for greater than	>	object **>** similar object	Boolean
test for greater than or equal	>=	object **>=** similar object	Boolean
logical and	and	logical array or Boolean **and** logical array or Boolean	Logical array or Boolean
logical or	or	logical array or Boolean **or** logical array or Boolean	Logical array or Boolean
logical complement of and	nand	logical array or Boolean **nand** logical array or Boolean	Logical array or Boolean
logical complement of or	nor	logical array or Boolean **nor** logical array or Boolean	Logical array or Boolean
logical exclusive or	xor	logical array or Boolean **xor** logical array or Boolean	Logical array or Boolean
logical complement of exclusive or	xnor	logical array or Boolean **xnor** logical array or Boolean	Logical array or Boolean

Details and examples

- Binary logical operators (and, nand, or, nor, xor, xnor)

Description: AND – when both arguments to the function are true, then the result is true.

OR – when either or both arguments to the function are true, then the result is true.

XOR – "exclusive" OR – when only either of the arguments to the function is true, then the result is true.

The "N" for NAND, NOR, and XNOR represents "NOT" and inverts the output of the AND, OR, and XOR functions.

Truth Tables for basic logical functions (where *X* and *Y* are inputs and *Z* is the output):

X	*Y*	*Z*
0	0	0
0	1	0
1	0	0
1	1	1

Table 3.3 - Truth Table for Function AND

X	*Y*	*Z*
0	0	1
0	1	1
1	0	1
1	1	0

Table 3.4 - Truth Table for Function NAND

X	*Y*	*Z*
0	0	0
0	1	1
1	0	1
1	1	1

Table 3.5 - Truth Table for Function OR

X	*Y*	*Z*
0	0	1
0	1	0
1	0	0
1	1	0

Table 3.6 - Truth Table for Function NOR

X	*Y*	*Z*
0	0	0
0	1	1
1	0	1
1	1	0

Table 3.7 - Truth Table for Function XOR

X	*Y*	*Z*
0	0	1
0	1	0
1	0	0
1	1	1

Table 3.8 - Truth Table for Function XNOR

Example: If you've defined:

```
signal single_value_a, single_value_b, single_result : std_logic;
signal vector_value_a, vector_value_b, vector_result: std_logic_vector(...);
```

Spec 3.5

then you may use the functions as follows:
single_result <= single_value_a **and** single_value_b;
vector_result <= vector_value_a **xnor** vector_value_b;

This illustrates that these functions can be applied to a single value (a "bit") such as a std_logic or to entire arrays such as a std_logic_vector where each member in the array is logically combined in the manner specified with the equivalent member in the other array.

A more detailed example based on the example immediately above...

```
single_value_a <= '1';
single_value_b <= '0';
single_reset <= single_value_a and single_value_b; - result will be '0'
vector_value_a <= X''FA'';
vector_value_b <= X''32'';
vector_result <= vector_value_a xnor vector_value_b; - result will be X''37''
```

Spec 3.6

Note that both vector_value_a and vector_value_b are the same length – VHDL won't fill the shorter vector to automatically match lengths with the longer vector.

What about logically combining a vector with a single bit such as:

"vector_result <= vector_value_a and single_value_a;" ?

This won't work either as the two operands must be of the same type.

Remember that VHDL is not case sensitive (the exception being the values for std_logic and std_logic_vector values), you may use **AND** or **and**.

Relational operators (=, /=, <, <=, >, >=)

Relational operators are used with if, while, and generate statements. Unlike C/C++ which uses a double equal sign for conditional equality, VHDL uses a single equal sign. You have already seen the signal operator ("<="), and you'll be introduced to the variable assignment operator (":=") shortly.

Testing magnitude for integers is a valid task. Fortunately this is well behaved and not noteworthy. Pay attention to how the conditional is constructed in this example and not the specifics of the syntax for the conditional assignment as this will be illustrated later.

```
...
   signal address_bus : integer range 0 to 65535;
...
begin
   ...
   Decode <= '0','1' when ((address_bus >= 0) and
                           (address_bus <= 1023));
   ...
```

Snippet 3.18

Testing std_logic_vectors for equality or inequality is meaningful and commonly performed. Testing std_logic_vectors for magnitude (greater-than or less-than) has no meaning as std_logic_vectors are just patterns and have no positional values. In fact, many compilers will throw an error indicating that this is an invalid comparison.

Best Design Practices 3.3 – Tip

Only test std_logic_vectors for equality or inequality!

If a std_logic_vector has a positional meaning, for example, representing an address on a bus, and it needs to be tested to see if it should be decoded, then the std_logic_vector should be cast to an unsigned or converted to an integer type. At this point it can be compared to the address range.

```
...
signal address_bus : std_logic_vector(15 downto 0);
...
begin
   ...
   decode <= '0','1' when ((unsigned(address_bus) >= 0) and
                          (unsigned(address_bus) <= 1023));
...
```

Snippet 3.19

Here is an excellent place to illustrate how the Booleans data type can be effectively used...

Example:

```
1: signal resultOfComparison: boolean;
2: ...
3: resultOfComparison <= (value1 > value2);
4: ...
5: if (resultOfComparison) then
6: ...
7: end if;
```

Snippet 3.20

Line 1: *resultOfComparison* is defined as a Boolean.

Line 3: *resultOfComparison* is assigned either **true** or **false** depending on the values of *value1* and *value2*.

Line 5: within a process (details to be discussed later). *resultOfComparison* is used to determine if the statements associated with the **true** condition are evaluated or not.

Shifting operators (sll, srl, sla, sra, rol, ror)

Shifting operators are a convenient way of rearranging data in hardware without the cumbersome use of array slices. Versions for both logical shifting and arithmetic shifting are available. These shifting operators are valid for the bit_vector type, but not intrinsically for the std_logic_vector type. In order to use these with the std_logic_vector type the IEEE library STD_LOGIC_1164 package must be loaded and the appropriate function called.

The following operators are valid for bit_vectors:

Table 3.9

Operation	Name	Description
sll	Shift left logical	Contents are shifted to the left. Leftmost bit drops into universal bit bucket. Rightmost bit filled with a '0'.
srl	Shift right logical	Contents are shifted to the right. Rightmost bit is lost. Leftmost bit filled with a '0'.
sla	Shift left arithmetic	Contents are shifted left. Rightmost bit is replicated. Leftmost bit is lost.
sra	Shift right arithmetic	Contents are shifted to the right. Leftmost bit is replicated. Rightmost bit is lost.
rol	Rotate left	Contents are shifted left. Leftmost bits are shifted into the rightmost positions.
ror	Rotate right	Contents are shifted right. Rightmost bits are shifted into the leftmost position.

Example:

```
signal shift_register : bit_vector(7 downto 0) := X"08"; ...
shift_register <= shift_register sll 4;
```

Results in shift register is X"80".

Signed and unsigned types are more commonly used. If an std_logic_vector needs to be shifted/rotated, it can easily be cast as an unsigned type, then shifted/rotated. We will cover costs and conversions shortly. Here are the shift/rotate functions defined in the IEEE NUMERIC_STD package:

Table 3.10

Function Name	Arg1	Arg2	Returns	Description
shift_left	unsigned	Natural	unsigned	Shifts the unsigned arg1, arg2 positions to the left
shift_right	signed	natural	Signed	As above
shift_left	unsigned	Natural	Unsigned	Shifts the unsigned arg1, arg2 positions to the right
shift_right	Signed	natural	Signed	As above
rotate_left	unsigned	Natural	unsigned	Rotates the unsigned arg1, arg2 positions to the left
rotate_left	signed	natural	Signed	As above
rotate_right	unsigned	Natural	Unsigned	Rotates the unsigned arg1, arg2 positions to the right
rotate_right	Signed	natural	Signed	As above

Some examples:

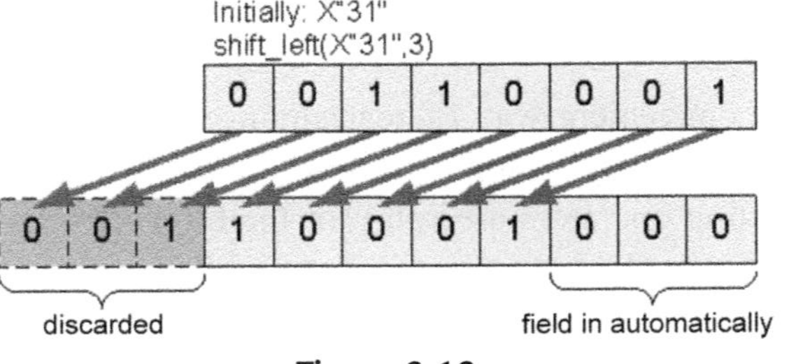

Figure 3.12

Resulting is an unsigned X"88".

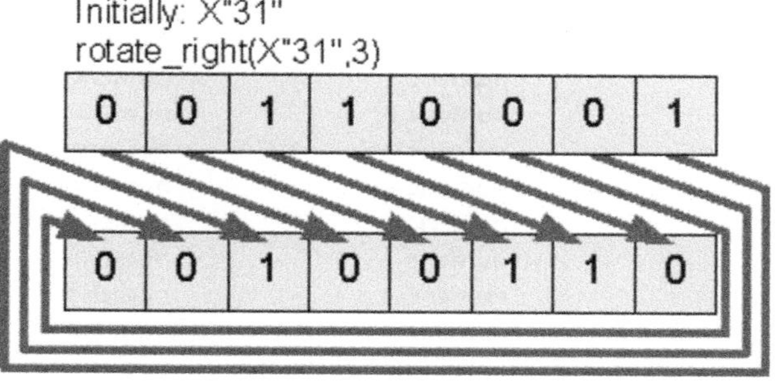

Figure 3.13

Resulting is an unsigned X"26".

Std_logic_vector is very commonly used and can be cast to use the unsigned shift functions; however, most find simple coding to be an effective and clear strategy for shifting. The '&' operator is a concatenation symbol that joins two std_logic/std_logic_vectors into a single std_logic_vector. The '&' operator can also be used to join strings and characters.

```
Line 1: signal slvValue : std_logic_vector (7 downto 0) := X"31";
Line 2: signal rotRightResult: std_logic_vector(7 downto 0);
Line 3: signal shftLeftResult: std_logic_vector(7 downto 0);
Line 4: …
Line 5: shftLeftResult <= slvValue(4 downto 0) & "000";
Line 6: rotRightResult <= slvValue(2 downto 0) &
                          slvValue(7 downto 3);
```

Snippet 3.21

Line 1: slvValue is defined as an 8-bit std_logic_vector initialized to hexadecimal 31.

Lines 2,3: result values are defined to be the same type and size as slvValue.

Line 5: the bottom 5 bits of the slvValue are preserved and three zeros are appended to the right. This has the effect of pushing the bottom 4 bits of slvValue to the left producing a left shift by three. The reason that the entire slvValue wasn't used is that slvValue has a size of 8, so when the additional 3 bits were appended, the net result would have been a vector of length 11. If shftLeftResult were defined to be an 11-bit vector, then this would have been a legal assignment.

Line 6: rotation can also be thought of as breaking the vector into two pieces and swapping. Since we are looking to do a right rotation of three, we take the lowest 3 bits and swap them with the upper 8–3 bits.

Mathematical operators: (+, −, *, /, mod, rem, &, not, abs, **)

VHDL provides the standard mathematical support – addition (+), subtraction (−), multiplication (*), and division (/) as well as the remainder function (rem), absolute value (abs), and modulo function (mod). These are defined for integers (synthesizable/simulateable) and reals (simulation only).

Table 3.11

Operator	Arg1 (left side)	Arg2 (right side)	Result	Description
+	Numeric	numeric	numeric	Addition
–	Numeric	numeric	numeric	Subtraction
*	Numeric	numeric	numeric	Multiplication
/	Numeric	numeric	numeric	Division
**	Numeric	numeric	numeric	Exponentiation
Rem	Numeric	numeric	numeric	Remainder
abs	Numeric	numeric	numeric	Absolute value
Mod	Numeric	numeric	numeric	Modulo
Not	Numeric	numeric	numeric	1's complement
&	Logic/logic_vector	Logic/logic_vector	Logic_vector	Concatenation

It is important to note that these functions may behave differently or have certain limitations when synthesized vs. when simulated. All simulation happens within the computer, therefore, all of these functions can be implemented in their full capacity. Sometimes the full capacity is limited such as with exponentiation which allows the left side of the exponentiation operator to be of any numeric type, but requires the exponent (right side of the exponentiation operator) must be an integer. This makes raising a real number to a real number impossible unless you write an overloading function and work the math yourself. Fortunately, the IEEE offers a collection of math libraries which can be found in the MATH_REAL package which provides a great deal of support for the real data type including logs, square roots, exponentiation, etc. This library is NOT synthesizable, but can be used in synthesizable code to produce constants.

```
signal rX, rZ : real;
signal exponent : integer range 0 to 10;
…
rX <= 2.1;
exponent  <= 3;
rZ <= rX ** exponent;   -- rZ will have the value of 9.261
```

Snippet 3.22

Synthesis places additional restrictions on what can be constructed internal to the FPGA. We know that the real type can't be implemented in synthesis, so the above example is clearly not synthesizable, but what about…

```
signal iX, iY, iZ : integer;
signal exponent : integer range 0 to 10;
…
iX <= 3;
iZ <= iX ** iY;
```

Snippet 3.23

Is this legal?

The simulation environment is perfectly happy about this, but in synthesis, this won't fly! Most synthesis tools restrict exponentiation to numbers which are powers of two (of which iX being 3 is not!). Additionally, some synthesis tools require both arguments of this operation to be constants! How can this be of any use?

As we will see in the final section of this book, code can (and should) be written in a fairly generic fashion so that it can be reused in other designs. The ability to compute values, even when they don't change when operating in the FPGA, is very useful as it means that reusable modules don't need to be hand edited for every instance which greatly reduces errors and frustration.

Concatenation

Moving away from numeric predefined math functions let's take a quick look at the catenation ('&') operator which is used to join bits, characters/strings. or std_logic values together to form vectors.

```
1: signal slv1      : std_logic_vector(15 downto 0)  := (others=>'0');
2: signal slv2      : std_logic_vector( 3 downto 0)  := X"7";
3: signal sl_value  : std_logic := '1';
4: signal slvResult : std_logic_vector(23 dowtno 0);
5: …
6: slvResult <= slv2 & slv1(10 downto 0) & X"F0" & sl_value;
```

Snippet 3.24

Line 1: slv1 is defined as a 16-bit std_logic_vector initialized to all '0's.

Line 2: slv2 is defined as a 4-bit std_logic_vector initialized to pattern 7 ("0111").

Line 3: sl_value is defined as a single value of type std_logic which is initialized to '1'.

Line 4: slvResult is defined as a 24-bit std_logic_value.

Line 6: slv2 and the lower 11 bits of slv1 are joined together to form a 15-bit std_logic_vector. The constant X"F" (4 bits of ones) is appended. Finally, the single bit of sl_value is appended in the rightmost position. Graphically:

slv2 & slv1(10 downto 0) & X"F" & sl_value;

"0111"& "000_0000_0000" & "1111" & '1'

Forming a result of "0111_0000_0000_0001_1111" or X"7001F"

Notice the following:

1. *Catenation occurs positionally, that is, the first element in the list of catenations appears at the leftmost position, and as additional elements are added, the values are added on the right.*
2. *The final value of the catenation operation must match the size and type of the variable or signal that is on the left-hand side of the assignment operator.*

From the example we can see that constants, std_logic_vectors, std_logic_vector slices, and std_logic types can be combined to form a single std_logic_vector provided that the requirements of item 2 are met.

Converting and casting

Casting

Casting is the process of reassigning a type without changing the underlying data structure. For example a std_logic_vector is merely an ordered collection of std_logic elements – the individual positions have no predefined meaning. A signed type (as defined in the numeric_std package) is also an ordered collection of std_logic elements, but in this case it is understood that the rightmost position represents the least significant bit, and the next to the leftmost position represents the most significant bit, with the leftmost position representing the sign. Now, with this information, mathematical and comparison functions can be performed.

Changing information from a std_logic_vector to a signed or unsigned is simple – it can be cast, or renamed, without impacting performance as it is a renaming of the positions which results in no additional hardware being generated.

Examples:

Table 3.12

std_logic_vector	Signed	Unsigned
X"F3" means "1111_0011"	X"F3" means −13	X"F3" means 243
integer	natural	positive
123	123 (but limited between 0 and INTEGER'HIGH)	123 (but limited between 1 and INTEGER'HIGH)

SIDEBAR 3.1

Why Is Casting and Conversion so Important?

vWe know that one of the most commonly used data types for FPGA design is that of the std_logic_vector, which is an array of the enumerated type std_logic. These types (std_logic/std_logic_vector) are the "bread-and-butter" of the FPGA designer. Data are most frequently moved in and out of the FPGA as a std_logic_vector, but represents some real quantity external to the FPGA. Consider an Analog-to-Digital converter. Although the signals arriving at the FPGA are likely to be represented as std_logic_vector, it is usually a good idea to get this representation into a more intuitive and usable form such as an unsigned or signed type, or an integer (unfortunately, physical types are not synthesizable either, even though the value from the AtoD represents a voltage or current, you can't use the physical data type in a synthesizable design - that said, these data types are very useful when building a board-level simulation). Recall that std_logic_vectors don't have any positional value; however, std_logic_vectors can be easily cast to an unsigned or signed data type. The IEEE library's NUMERIC_STD package includes overloading functions that allow the basic operations to be used on signed and unsigned types, but not on std_logic_vectors directly.

Casting examples:

```
1: unsigned_result <= slvValue1 mod 4;
2: unsigned_result <= unsigned(slvValue1) mod 4;
3: integer result <= integer(slvValue);
```

Snippet 3.25

Line 1: illegal – slvValue1 (as designated by its name) is a std_logic_vector. As we've mentioned before, there is no positional meaning to a std_logic_vector, therefore performing any type of math operation is meaningless.

Line 2: in this case, we've "cast" the std_logic_vector to an unsigned – no change of bits has occurred, but the unsigned data type places the least significant bit on the rightmost side and assigns it the positional value of one. Now that this is a valid "number" the modulo function can be performed on it.

For synthesis, the modulo function may also have further restrictions such as requiring the second operand to be a power of 2 (which is then easy to implement as the appropriate number of upper bits are simply forced to ground).

Line 3: illegal – here we see an attempt to cast a std_logic_vector to an integer. Because this is a more involved process, this must use a conversion function.

This type of behavior between the simulation capabilities of a function and the synthesizable capabilities of a function is seen throughout the mathematical functions.

A list of functions supported by the NUMERIC_STD package can be found in the NUMERIC_STD section of the IEEE library description toward the end of this book.

Conversion functions

Conversion functions, like casts, change the way information is stored or the way that it is represented. Conversion functions may also require more information than a simple cast.

Consider converting an integer into a std_logic_vector. If we want to change an integer into a std_logic_vector we can't simply make this conversion unless we wrote our own conversion function. If we are to use the "standard" conversion functions provided in the numeric_std package, we would first need to convert the integer to a signed or unsigned, then cast it to a std_logic_vector. Would we be done here? No! There is one more piece of information needed – how big should the std_logic_vector be? 32 bits, 64 bits, 4 bits? You must provide this information to the conversion function.

The to_signed/to_unsigned conversion functions provided in the numeric_std package requires us to provide not only the integer to convert, but also the size of the target signed/unsigned.

```
1: signal slvValue1  : std_logic_vector(31 downto 0)  :=
                       (11 downto 8 | 1 | 0 =>'1', others=>'0');
2: signal slvValue2: std_logic_vector(31 downto 0)  :=
                       (others=>'U');
3: signal usValue   : unsigned(31 downto 0)  := (others=>'0');
4: signal iValue    : integer range -8192 to 8192 := 0;
5: . . .
6: usValue    <= unsigned(slvValue1);
7: iValue     <= to_integer(usValue);
8: slvValue2 <= std_logic_vector (to_signed(iValue,32));
```

Snippet 3.26: Examples of Casting and Conversion Functions.

Code analysis:

Line 1: signal *slvValue1* is set to be a std_logic_vector of 32 bits with a hexadecimal pattern of X"0000_0F03".

Line 6: signal *slvValue* is cast to an unsigned type and placed in signal *usValue*. The bit pattern remains as X"0000_0F03".

Line 7: the conversion function provided in the IEEE numeric_std package is utilized to change the unsigned signal *usValue* into an integer signal *iValue*. While it is possible that the bit pattern is unchanged, this is not a certainty and the specific pattern used to hold integers (as well as reals and other types) is up to the tool vendor.

Line 8: shows the nesting of both a conversion function and a cast. First the signal *iValue* is converted into a 32-bit signed temporary value, then cast to a std_logic_vector. At the point that the signal *iValue* has been converted, the bit pattern returns to X"0000_0F03".

Since the ability to derive an address bus width for a given block of memory is a fairly common task, let's look at a simple example here. Note that the mathematical operations are performed completely during synthesis and result in a static (constant) value during the FPGA's operation. Although this process occurs for synthesis, it is not considered synthesizable as it does not produce gates in the FPGA.

```
1: constant memory_size : integer := 32768;
2: constant bus_width    : integer :=
  integer((LOG(real(memory_size))/LOG(2.0)+0.99999);
3: signal   address_bus : std_logic_vector(bus_width-1 downto 0);
```

Snippet 3.27: Use of Casts in a "Real-World" Application.

Code analysis:

Line 1: the size (depth) of the memory block is specified.

Line 2: based on the size of the memory, the next largest power of 2 boundary is found. Note that the *memory_size* is first cast to a real as this is the argument required by the LOG function in the MATH_REAL package. Once the value is computed, then it is set to the next value higher (so that remainders are detected).

Line 3: finally the integer value is used to set the proper size of the address bus.

3.1.7 Concurrent Statements

Concurrency is often one of the more difficult concepts for software folks to grasp when transitioning into hardware. Traditional computer systems execute a single instruction at a time and only a single instruction. Mechanisms such as context switching (task swapping) give the appearance of concurrency, but this is only an appearance and not an actuality. There will be those that point out (and correctly so) that multiple cores and multiple processors do offer true concurrency; however, this is nothing compared to what happens inside an FPGA!

Remember what we're doing when we code in VHDL – we are describing an actual logic circuit to be implemented in hardware, so concurrency is quite real.

For the purposes of building a mental model of what's happening, let's use an analogy of a pipe and valve. The pipe represents the wire on which a single bit of a signal travels. The valve represents a simple flip-flop (see Figure 3.14). Finally, the datum itself (the voltage) is represented by a marble for logic high or not a marble (the absence of a marble) to represent a logic low.

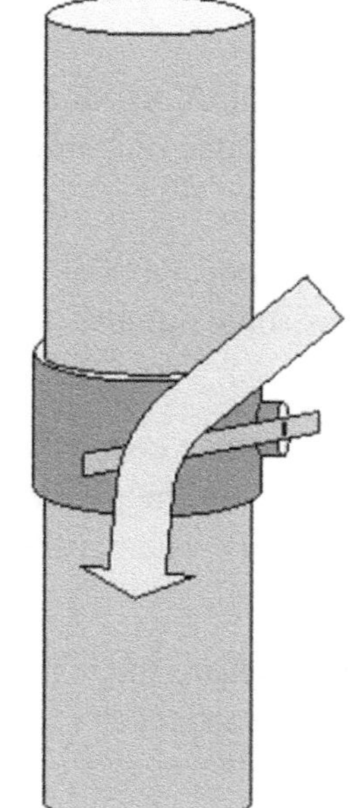

Figure 3.14: Single "wire" with "flip-flop".

For the purposes of this exercise, the marbles represent bits and the track represents routing resources. Every rising edge of the clock (one clock period), the handle on the valve is rotated vertically allowing the marble or absence of marble to pass. (Using this model, the set-up and hold time corresponds to how long it takes for the marble to initially bounce off the stopper and actually come to relative rest. The clock-to-out timing is the time it takes to open the valve to let the marble pass.)

The "simultaneity" comes from multiple tracks with multiple balls ("bits") moving from place to place (see Figure 3.15).

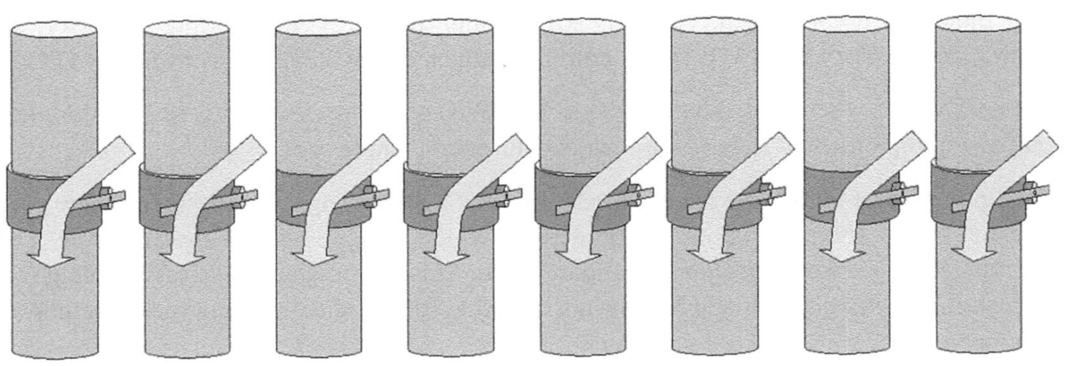

Figure 3.15: With concurrency, information moves in multiple paths simultaneously.

Not every signal is going to move from point 'A' to point 'B' and each valve is not necessarily going to be opened and closed at the exact same time or rate. A slightly more realistic view of this model might be seen in Figure 3.16.

Combinatorial logic is harder to build a physical model of, but for the time being let's just say that there's a magic black box that takes some number of marbles in and produces (or doesn't produce) a marble out.

Using this model, the "plan" of how each pipe connects with logic functions represents the VHDL source code and the task of laying out the pipe and function boxes represents the process of implementation (see Figure 3.17). The marbles moving through the pipes are the runtime events.

Ok, and now picture the whole assembly where the lengths of the pipe are now measured in dozens of nanometer, there are many thousands if not millions of pipes, and the marbles are moving just under the speed of light…

This kind of concurrency results in tasks being performed in parallel which yields much higher performance than a fixed-architecture processor. The challenge here becomes one of

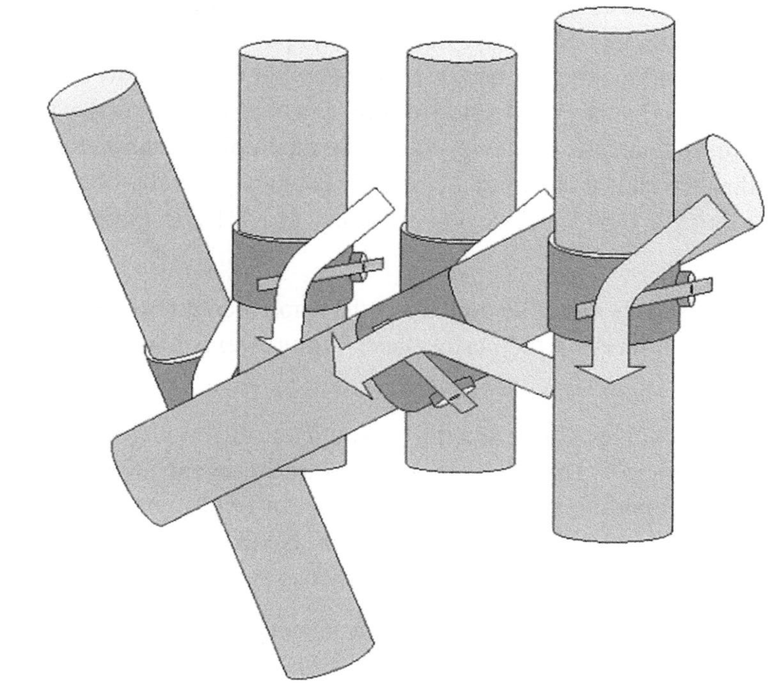

Figure 3.16: Not every pipe goes to the same place or has its valve opened at the same time.

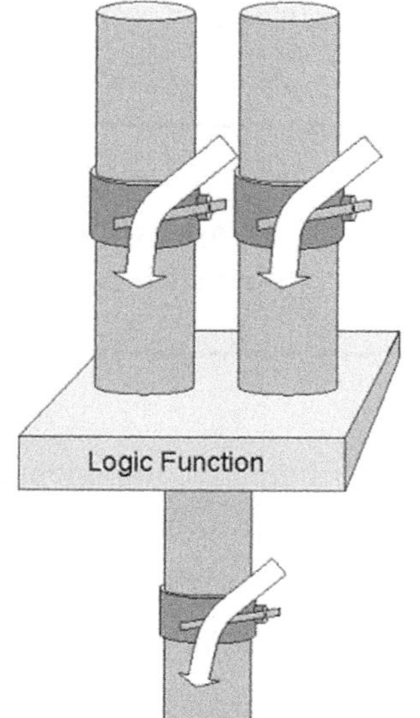

Figure 3.17: Logic function producing output based on multiple inputs.

synchronizing information as it flows through the FPGA. This synchronization falls into two categories – making certain that data arrive at the correct point simultaneously and quickly enough (performance), and making certain that the logic that the data flow through does not excessively delay the data (latency). “Meeting timing” is a phrase meaning that the performance goals have been achieved, and the processes of driving the implementation tools (and occasionally recoding) is known as “timing closure”. Latency generally becomes less of an issue the faster the clock is run (1 clock cycle of latency at 1 MHz is longer than 3 clocks at 100 MHz).

Technically, timing closure is more the domain of the implementation tools; however, if poor coding style is used, then meeting the performance requirements can be difficult if not impossible.

Consider: we’ve added “valves” quite liberally throughout this model – this results in well-timed and high-performing FPGAs. Unlike the hardware store, each length of pipe (technically, each logic function) contains its own valve for free, so using valves doesn’t cost any more than not using valves. If one thinks of the FPGA logic in this fashion, then synchronous, high-performing designs are much easier to realize.

Think about how the system might behave if there were very few (or even no) valves: marbles would arrive at logic functions delayed only by the length of pipe they traveled. Some marbles might travel a very short distance while others might travel a much longer distance. Suddenly you realize that the marbles aren’t going to arrive at the same place at the same time (think race conditions) and after only a few logic functions, the marbles have to be dropped so infrequently into the pipe that the performance becomes intolerable!

Best Design Practices 3.4 – Tip

Use registers (valves) freely to synchronize the flow of information in the FPGA!
Corollary: This may increase the latency of the system, however, with increased clock speeds this latency is minimized.

During the course of this book, you will create components – roughly equivalent to “functions” or “methods” in C/C++ and Java, processes, and individual concurrent statements – all of which will be operating at the same time instead of “taking turns”.

Now that you have a “feel” for what concurrency is, let’s look at how to implement it…

Concurrent statements can take one of several forms for synthesis: assignment, process, instantiation. Several other concurrent statements exist for simulation which we will discuss more when we cover test benches.

Unconditional assignment (<=)

Concurrent assignment statements can be unconditional (absolute), where the operation described on the right-hand side of the assignment statement is assigned to the signal on the left-hand side with no delay, always.

There are some basic rules for performing assignments in VHDL. They are as follows:

1. Both sides of the assignment operator must be of the same type, that is, if the signal on the left-hand side of the operator is a std_logic, then the result of the right-hand side of the operator must produce a std_logic result.
2. Both sides of the assignment operator must be of the same size. For example, if a std_logic_vector(3 downto 0) is on the left-hand side of the operator, then the right-hand side of the operator must produce a 4-bit std_logic_vector.
   a. Some VHDL compilers require that the direction of the vector must be the same, so if the left-hand signal was defined using a "downto" statement, then the right-hand side must produce a result that is in the same direction.

Consider the following examples:

```
 1: signal x,y,z   : std_logic;
 2: signal zzz     : std_logic_vector(1 downto 0);
 3: signal a       : integer range 0 to 100;
 4: signal b       : bit;
 5: …
 6: z <= x;
 7: z <= x AND y;
 8: zzz <= x AND y;
 9: zzz <= x & y;
10: z <= a;
11: z <= b;
12: zzz <= 3;
13: A <= x + y;
```

Snippet 3.28

Line 1: *x*, *y*, and *z* are defined as std_logic (which is 1 bit wide).

Line 2: zzz is defined as a std_logic_vector of width 2.

Line 3: *a* is defined as an integer which may have a value between 0 and 100.

Line 4: *b* is defined as a bit type (width 1).

Line 6: OK, both *z* and *x* are of the same type and size.

Line 7: OK, *x* AND *y* produce a result compatible in both size and type as *z*.

Line 8: Illegal: zzz is defined as a std_logic_vector of width 2, but *x* AND *y* produce a std_logic result.

Line 9: OK, *x* & *y* produce a 2-bit std_logic_vector result.

Line 10: Illegal: *z* and *a* are of different types.

Line 11: Illegal, even though *z* and *b* represent items that are 1-bit wide, a std_logic type can represent nine different values whereas a bit only represents two values therefore these are incompatible. You might rationalize this by thinking that bit has '1' and '0', and so does std_logic, but this is incorrect thinking. '1' and '0' may mean something completely different for bit than in std_logic (in this usage, it doesn't, but the synthesis tools don't know that). A conversion function is required to make this assignment valid.

Line 12: Illegal: although 3 can be represented as 2 bits which would fit into zzz, std_logic_vectors have no intrinsic positional values associated with them therefore this is essentially attempting to force an integer into a std_logic_vector – an illegal operation. This could be resolved easily by either representing 3 as "11" or converting 3 as a signed or unsigned type, then casting as a std_logic_vector.

Line 13: Illegal. Using "straight" VHDL this is illegal as the '+' operator is only defined for integer types. You could write an overload function for the '+' operation that would make this legal.

Where can assignments be performed? Assignments (to signals) may occur both inside and outside of a process (which we're coming to soon), and only between the beginning and end statements of an architecture. It is important to note that assignments behave slightly differently when appearing inside a process block than when appearing outside a process block. This will be thoroughly covered when processes are discussed.

Conditional assignment – when/else

All the rules for assignments are the same as described above, but now only occur when the conditional evaluates to logic true. If the conditional evaluates to logic false, then if no else clause is present, no assignment is made. If the else clause is present, the "false/else value" is assigned to the target.

There are two basic forms of syntax for a conditional assignment. The first is informally known as the "when/else" form and is most similar to the "if/then/else" type of operation from software. You might ask why VHDL doesn't implement an if/then/else – in fact it does, but there are some restrictions on it which we will discuss in the process section.

In the meantime, here are some simple examples of the when/else concurrent conditional assignment statement:

```
target <= thing1 when (conditional);
target <= thing1 when (conditional) else thing2;
target <= thing1 when (conditional1) else
          thing2 when (conditional2) else
          ...
          thingn;
```

Snippet 3.29

Note: These three statements are not intended to be found in the same body of code as will essentially "short-out". Fortunately this will be flagged as a multi-source error.

Line 1: represents the simplest form of when/else statement. When the conditional is met, thing1 is assigned to target. Until condition is met, target remains in its power-up state. This is roughly equivalent to an *if (conditional) then assign end if* type of statement from software.

Line 2: as with Line 1, thing1 is assigned to target when the specified conditional is met. If the conditional evaluates to false, then thing2 is assigned to target. The software equivalent to this statement is "if/then/else".

Line 3: The most complex form of this statement assigns thing1 to target when conditional1 is met, otherwise, if conditional2 is met, then thing2 is assigned to target, and so forth. If none of the conditionals evaluate to true, then thingn is assigned to target. Note that the final else and thingn can be omitted so that until any of the conditional statements are met, target would retain its power-up value.

This structure is convenient to use when any of the conditional statements may overlap or contain different values for comparison. The order of the testing of the conditionals determines the priority of assignment.

Common use examples:

```
1: syncdSignal <= someSignal when (rising_edge(clk));
2: muxSelect   <= '1' when (code = X"13") else '0';
3: resetSystem <= '1' when (externalReset = '1') else
4:                '1' when (cpuReset = '1') else
5:                '1' when (otherReset_n = '0') else
6:                '0';
```

Snippet 3.30

Line 1: syncdSignal is assigned the value of someSignal when the clk signal transitions from low to high. You must make certain that someSignal is already synchronized to clk otherwise you run the risk of meta-stability in syncdSignal. This style of coding will typically infer a rising edge sensitive D-type flip-flop with *someSignal* as the D input, *clk* being the clock input, and *syncdSignal* being the Q output.

Line 2: muxSelect is assigned a logic high when code corresponds to the pattern "0001_0011", otherwise it is forced to logic low.

Line 3: this line could just as easily be rewritten as "resetSystem <= externalReset OR cpuReset or not otherReset_n; however, this form is especially maintainable as each specific condition is spelled out and the inevitable error due to the use of negative logic is avoided. Notice that none of the conditionals used in this line share the same signals. This is one area where this type of conditional assignment is superior to the with/select form shown in the next section.

Since the conditionals can be completely independent of one another, there is always the possibility of coding an overlapping condition. In this case, the conditional that is tested earlier (closer to the assignment) is given priority over subsequent conditional assignments. For example:

```
target <= UPPER_RIGHT when ((angle >=   0) and (angle <=  90)) else
          LOWER_RIGHT when ((angle >=  90) and (angle <= 180)) else
          LOWER_LEFT  when ((angle >= 180) and (angle <= 270)) else
          UPPER_LEFT  when ((angle >= 270) and (angle <= 360)) else
          BULLSEYE    when (radius < 10);
```

Snippet 3.31

Assume that UPPER_RIGHT, LOWER_RIGHT, LOWER_LEFT, UPPER_LEFT, and BULLSEYE are defined as an enumerated type and that target is of the same enumerated type. Angle is an integer with a subrange of 0 to 360 and BULLSEYE is an integer with a subrange of 0 to 100.

It is easy to see that if the angle has a value of 0 to 89 inclusive, then target will be assigned a value of UPPER_RIGHT, but what happens when angle is 90? Closer inspection shows that both UPPER_RIGHT and LOWER_RIGHT should decode in this case; however, because UPPER_RIGHT is evaluated before LOWER_RIGHT, then target will be assigned UPPER_RIGHT.

Now that this situation is observed, we see it repeat with each of the corner cases – 180 and 270; 360 is a special case in that it is functionally equal to 0, but because it's unique in the conditional statements, it will decode separately (somebody will have to have a chat with the designer of the angle measurement, though).

Finally, when will BULLSEYE be assigned to target? Never! Since angle is limited to a value between 0 and 360 and the four conditionals cover all possibilities, then regardless of any value that radius takes on, target will never receive the value of BULLSEYE.

Multiple selection conditional assignment (with/select)

The second type of conditional assignment that VHDL offers provides a mechanism to perform multiple conditions resulting in a single assignment. This is roughly equivalent to a switch or case statement.

This conditional assignment differs from the with/select form in that a selection is made based on the value of only one signal (in the example below, the signal *selector*). This type of structure prohibits overlapping conditions. If not every possibility is covered in the selection list then a default value can be assigned using the "when others" clause as shown below:

```
with (selector) select
    target <= thing1 when (value1),
              thing2 when (value2),
              thing3 when (value3),
              thing4 when others;
```

Snippet 3.32

Line 1: signal *selector* is identified to be the signal against which each condition is tested.

Line 2: target is assigned thing1 when *value1* matches selector. As always, the type and size of *valueX* must match that of *selector*.

Line 3: should *value1* not match the current value of *selector*, then *value2* is checked. If it matches, then *thing2* is assigned to *target*.

Line 4: the process continues for as many or few possible values as are needed to be checked.

Line 5: should none of the previously identified values match that of *selector* then *thing4* is assigned to *target* due to the use of the "when others" clause.

Let's continue the example from the "when/else" (Snippet 3.31) example. Let's assume that we need to drive specific non-overlapping std_logic_vectors based on the result of the quadrant that the angle falls into. The "when/else" example resolves the angle into quadrants, and now that all of the conditions are the same (the result is always going to be of type QUADRANT), we can use the with/select statement to select which std_logic_vector pattern will be placed on the *sigOut* signal.

```
1: with (target) select
2:     sigOut <= X"AAA" when (UPPER_RIGHT),
3:               X"555" when (LOWER_RIGHT),
4:               X"861" when (LOWER_LEFT),
5:               X"79E" when (UPPER_LEFT),
6:               X"606" when (BULLSEYE);
```

Snippet 3.33

Line 1: target, which is an enumerated type, is identified as the signal on which to test the various conditions against.

Line 2: sigOut, a std_logic_vector is assigned the pattern "1010_1010_1010" when *target* matches the member of the enumerated type, UPPER_RIGHT.

Line 3: should *target* not match UPPER_RIGHT, then it is tested against the enumerated type LOWER_RIGHT. Should they match, then the pattern "0101_0101_0101" is assigned to sigOut.

Line 4: *target* is tested next against LOWER_LEFT and if it matches then hexadecimal pattern X"861" is applied to sigOut.

Line 5: UPPER_LEFT is compared to *target* and if they match, then hexadecimal X"79E" is applied to *target*.

Line 6: finally *target* is tested against BULLSEYE and if it matches, then sigOut get pattern X"606". Noting that BULLSEYE can never be reached in the when/else test clause, it is likely to be optimized out.

What happens if *target* is of the proper enumerated type, but is not listed in this with/select statement? Should this happen, sigOut should not change value which implies that its previous value must be retained. In the next section we will discuss why this type of coding produces suboptimal performance.

The best way to avoid the above-mentioned problem is to explicitly declare a "catch-all" case using the "when others" clause. The previous example is better coded as:

```
1: with (target) select
2:     sigOut <= X"AAA" when (UPPER_RIGHT),
3:               X"555" when (LOWER_RIGHT),
4:               X"861" when (LOWER_LEFT),
5:               X"79E" when (UPPER_LEFT),
6:               X"606" when (BULLSEYE),
7:               X"000" when others;
```

Snippet 3.34

Lines 1–6: as described above.

Line 7: should no other conditional be found to match *target*, then sigOut will be assigned a pattern of all zeros. This extra "catch-all" avoids the issue known as "unintentional or accidental latch inference". We'll discuss inference right after this pearl of wisdom:

Best Design Practices 3.6 – Tip

Always use a "catch-all" condition to avoid accidental latch inference.

1. Use a final "else" when coding with the "when/else" construct.
2. Use a final "when others" when coding with the "with/select" construct.

Conditional assignment – time-based

Sometimes it is necessary to specify time as a criterion for making an assignment. We know that the *time* type is not synthesizable, however, in simulation it is very convenient to specify when something happens.

Let's say that we are modeling a reset signal which needs to remain active for 1 μs, then de-assert for the remainder of the simulation. This is easily accomplished with the following code:

```
reset <= '1', '0' after 1 us;
```

Snippet 3.35: Timed Association.

This line of code reads: reset is assigned the value of '1' until 1 μs has elapsed at which point reset takes on the value of '0'.

This structure can be extended to create more sophisticated waveforms such as:

```
parallelDataIn <= X"30", X"31" after 100 us, X"32" after 200 us;
```

Snippet 3.36: Multiple-Timed Associations.

This line of code reads: parallelDataIn is initially given the value of X"30"; 100 μs after the simulation has begun, parallelDataIn is then given the value of X"31". Finally, 200 μs after the simulation has begun or 100 μs after parallelDataIn has taken on the value of X"31", it takes on its final value of X"32" and holds this value until the end of simulation.

So, two important items to bear in mind when using this type of construct:

1. This is only valid in simulation.
2. The time given is measured from the beginning of simulation, not from the end of the previous period.

There is another common use for this type of construct – building periodic signals. Assume that you need to build a clock source with a 50% duty cycle with a period of *clock_period*, then you could code the following:

```
1: constant clock_period : time := 10 ns;      -- 100MHz
2: signal   clk          : std_logic := '0';   -- '1' or '0'
3: . . .
4: clk <= not clk after clock_period / 2;
```

Snippet 3.37: Building a Periodic Clock.

Line 1: a constant named *clock_period* is declared to be of physical type time and is initialized to 10 ns which corresponds to 100 MHz.

Line 2: a signal named clk is defined as a std_logic. Often signals are initialized to 'U' so that it is easy to identify when the signal has not been properly initialized. In this case, we must initialize this signal to a '1' or '0' so that the "not" operator on line 4 will be able to generate the inverse value. If this is left as uninitialized, then the operation for not 'U' is still 'U' and no oscillation occurs.

Line 4: clk is assigned the inverse of itself after half a clock period, which produces a 50% duty cycle at the specified frequency.

Because the signal clk is based on a function of itself, the delay "after clock_period/2" isn't measured from the beginning of simulation time, but rather from the last change.

Instantiations, inferences, and hierarchy

So far, we've discussed signals, datatypes, some data manipulation capabilities, and some conditional processing, all within a single layer of hierarchy. Now, we'll introduce three concepts critical to your understanding of VHDL – Hierarchy, Instantiations, and Inferences. Let's deal with the concept of Hierarchy first.

As we pointed out so many pages ago, the Entity statement describes, among other things, all of the signals that enter and exit this module. At the top level of the hierarchical design, these signals will become pins on the FPGA. If, however, we were to use this module as a piece of a larger design within the FPGA, then these signals would become access points in and out of this module. Figure 3.18 shows a higher-level module instantiating two lower-level modules. The lower-level modules connect to the higher-level module through the ports of the entity.

The higher-level module (see Snippet 3.38) must first define the module to be instantiated, then it may instantiate each one of these modules as many times as required.

Module definition can occur in one of two places: between the "architecture" statement and the "begin" statement or within a package which we will defer to the last loop in this book.

Remember that component definitions don't actually *do* anything – they just describe what this module may be connecting to (just because you define it, doesn't mean that you have to use it, although this isn't good design practice as this will just clutter your code). This means that if module_Alpha only needs to be defined once even if it is instantiated once, a dozen, or a thousand times within this module.

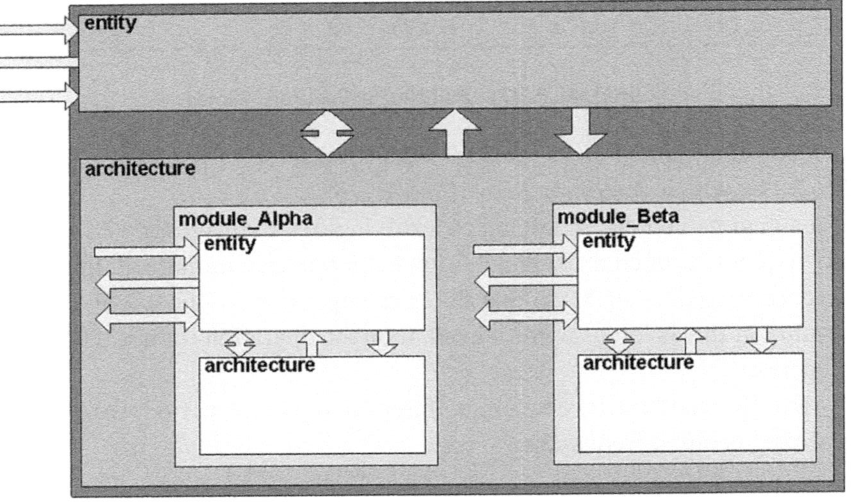

Figure 3.18: Relationship of "higher" level module instantiating 2 sub-modules[10].

[10] For illustrative purposes signals entering the entity can be inputs, outputs, or bidirectional (inouts). The source code illustration just shows simple inputs and outputs. There are a few other directional modes, however, they are seldom used when developing FPGAs and are omitted here.

Here's a fairly generic example of relationship of "higher" level module instantiating two submodules (see Snippets 3.38 and 3.39).

```
 1: entity higherLevelModule is
 2:   Port ( x1  : in  STD_LOGIC;
 3:          x2  : in  STD_LOGIC;
 4:          y1  : in  STD_LOGIC;
 5:          y2  : in  STD_LOGIC;
 6:          z   : out STD_LOGIC);
 7: end entity higherLevelModule;
 8:
 9: architecture Behavioral of higherLevelModule is
10:
11:    component module_Alpha is
12:       port ( a1 : in  STD_LOGIC;
13:              a2 : in  STD_LOGIC;
14:              a3 : out STD_LOGIC);
15:    end component module_Alpha;
16:
17:    component module_Beta is
18:       port ( b1 : in  STD_LOGIC;
19:              b2 : in  STD_LOGIC;
20:              b3 : out STD_LOGIC);
21:    end component module_Beta;
22:
23:    signal z1, z2 : std_logic := 'U';  -- connecting signals
24:
25:   begin
26:
27:         subA: module_Alpha port map (a1=>x1, a2=>x2, a3=>z1);
28:         subB: module_Beta  port map (b1=>y1, b2=>y2, b3=>z2);
29:         z <= z1 xor z2;
30:
31:   end Behavioral;
```

Snippet 3.38: Higher-Level Module Instantiating Two Lower-Level Modules.

Line 1: beginning of the entity description for the module *higherLevelModule.*

Lines 2–6: port definitions – x1, x2, y1, y2, *z* are all defined as signals of std_logic. All but signal *z* are defined as signals entering the module, which means that you cannot assign values to these signals, but rather treat them as constants; whereas signal *z* is defined as a module output which means it can only have other signals or constants applied to it – it cannot be used as an input.

Line 7: end of the entity statement.

Line 9: beginning of the architecture to pair with entity *higherLevelModule*. This architecture is named *behavioral.*

Lines 11–15: component declaration for module_Alpha. This component declaration must match the entity statement for module_Alpha. As mentioned before, the component declaration doesn't "do" anything, but serves as a template so that each time module_Alpha is instantiated, the tools "know" what type and size each signal is supposed to be, whether it is an input or output, and if all the signals are present.

Lines 17–21: component declaration for module_Beta.

```
entity module_Alpha is
   port ( a1 : in  STD_LOGIC;
          a2 : in  STD_LOGIC;
          a3 : out STD_LOGIC);
end entity module_Alpha;

architecture Behavioral of module_Alpha is
   begin
      a3 <= a1 and a2;
   end Behavioral;
```

```
entity module_Beta is
    Port ( b1 : in  STD_LOGIC;
           b2 : in  STD_LOGIC;
           b3 : out STD_LOGIC);
end entity module_Beta;

architecture Behavioral of module_Beta is
   begin
      b3 <= b1 or b2;
   end Behavioral;
```

Snippet 3.39: Source Code for Lower-Level Modules.

Line 23: signal definition for z1 and z2.

Lines 27 and 28 do the actual instantiation. Each instantiation must have a label (which is a good thing as it provides a reference back into your code when you use the simulation tool and many other analysis tools). The instantiation then describes the module that is to be instantiated, followed by the keywords "port map". We'll describe what this means in painful detail later, but for now take them to mean "the ports that are defined in the module to be instantiated entity list connect to the following signals from this module". This is then immediately followed by a mapping between the signals listed in the entity to be instantiated entity list (the name on the left-hand side of the "=>" symbol) and the signal in this module (the name on the right-hand side of the "=>" symbol).[11]

Important note: the "=>" is confusing as it seems to suggest a direction for connection. It does NOT! This symbol indicates an association.

So, line 27 instantiates a module named module_Alpha and connects the signal x1 from this module with the signal within module_Alpha named a1. Similar connections are made for x2 to a2, and z1 with a3. Note that these are connections – just like soldering two wires together. It is the designer's responsibility to make sure that input signals are properly driven. Output

[11] There is also a mechanism that makes associations by the position (i.e., order) of the arguments as opposed to by naming. Positional assignment associates the first item listed with the first element in the entity list, etc. I STRONGLY recommend AGAINST using this as the likelihood of introducing an error is extremely high. It also increases the difficulty in reading the code, and hence in its maintainability.

signals may be left disconnected (using the keyword "open") or tied to a signal which is not driven otherwise a "multi-source" error will occur (a short).

One last important point – just as with signal assignments, each signal must be of the same type and size when connecting to a submodule.

Instantiation and Inference

Instantiation is the EXPLICIT description of a lower-level module. As we saw above, code Snippet 3.39 defines, then instantiates module_Alpha. When we instantiate a submodule, we are coding in the "structural" style, that is, we are describing how various modules connect. The modules themselves may be coded in any fashion.

Inference is a suggestion to the tools as to how to construct a specific aspect of the design.

Consider: your boss mentions that it would be nice for the project to have a given feature – you ***infer*** that he wants you to go and build it the way you think it should be done. This is what the tools do – given light direction as to what to build, but not the specific details of how it should be accomplished, the tools, based on templates within the synthesis tool, decide how to build what you outlined in your code. Does this mean that you might build something that does what your boss wanted, but not exactly the way he wanted it? Sure, but it also means that you might have exceeded your boss's expectations too. This is the risk of inference. The tools will look at your code and infer that you want something built. Then it will optimize like crazy, usually producing a result in a different way than you expect, but it works! Sometimes, loose parameters can be specified with an inference – many synthesis tools will make certain tradeoffs as to how something is built based on synthesis tool options such as build for speed (which usually requires more resources, but is more likely to meet performance goals), or for area (which attempts to use as few resources as possible, but performance drops), or one of several other directives.

Your boss could also have said – use thus and such module from a library which uses a specific and immutable process. Now your task is straightforward, you only need to infer the module and connect it, but the tools may be deprived of their opportunity to optimize the code. Sometimes instantiations make use of dedicated silicon resources (a good thing) but you may be required to migrate the design to another family which may not have that exact silicon resource. Now recoding is required where it might not have been necessary if inference had been used.

Let's look at inference and instantiation more closely…

Suppose you need a simple D-type flip-flop with no set or reset, only an enable. You could have the tools infer this D-type flip-flop by coding:

```
Q <= D when (rising_edge(clk) and (enable = '1'));
```

Snippet 3.40

You could also have made use of a vendor's library. For example, Xilinx provides the UNISIM library which contains a macro for instantiating this type of flip-flop:

```
instFlipFlop: FDE port map (D=>D, C=>clk, CE=>enable, Q=>Q);
```

Snippet 3.41

The syntax is interesting on this statement as it has a number of important parts. First, the label "FlipFlop" is required to set this component apart from other components. In truth, it should be named something a little more meaningful such as "delay_sig_D" if this device is being used to delay signal D by one clock. FDE represents the name of the component which is being instantiated. Xilinx uses a standard naming convention so that the designer can tell exactly what parameters are required for a given flop. In this case 'F' represents a flip-flop, 'D' indicates that it is a 'D-type', and the 'E' shows that it uses an enable. The keyword "port map" indicates that the following list is going to map signals at this level of hierarchy with the port signals defined in the entity for the FDE component. This is an illustration of "named association" where each port listed in the components' port list is explicitly named and mapped to a signal at this level of hierarchy. The port signal in the FDE component named 'D' is wired to a signal named 'D' at this level of the hierarchy. Notice that the => operator shows association, not flow of information. This can be quite confusing for the first-time designer.

This naming process continues associating the port in FDE named 'C' with the signal clk; the port named 'CE' with the signal named enable; and the output 'Q' with the signal Q. Again, one would hope to use more meaningful names than D, enable, and Q, however, for this example, the illustration is that of a flop which has a fairly standard naming pattern.

There are a few advantages to using named association including the ability to order the list in any fashion. The following is functionally identical to the instFlipFlop example:

```
instFlipFlop: FDE port map (Q=>Q , C=>clk, D=>D, CE=>enable);
```

Snippet 3.42

Often designers will write out an instantiation on multiple lines which makes it much more readable when instantiating modules with many signals:

```
instFlipFlop: FDE port map (D  => D,          -- input to the F/F
                            C  => clk,        -- xfer: D to Q on rising edge
                            CE => enable,     -- F/F enable active high
                            Q  => Q           -- F/F output
                           );
```

Snippet 3.43

There is also a way to instantiate a module using "positional association". Instead of listing the signals in the module and making explicit connections, these connections are made by matching the order that they are declared in the component, such as:

```
instFlipFlop: FDE port map (D,clk,enable,Q);
```

Snippet 3.44

This is extremely prone to human error and may be difficult to debug.

Best Design Practices 3.7 – Tip

Always instantiate using named association!

Instantiation is also the technique that you will use to construct hierarchies within your design. Remember our old maxim from earlier "Design from top downward, build from the bottom up". This means that as you begin to code, start with the "leaf" modules of the design – the ones deepest in the hierarchy. These are usually fairly small and easy to implement. Once you've built the entity and its associated architecture (and, of course, after you simulate the module and verify its functionality), you can then combine this newly coded module with the other modules from this level of hierarchy into another module. By including other modules, you've begun to define a hierarchy.

Let's walk through a more realistic example: Suppose you would like to build an RS-232 Universal Asynchronous Transceiver (UART). Begin the design by outlining the top level of the UART, which may not necessarily be the top level of your design. We start with the block diagram for the top of the UART design (see Figure 3.19).

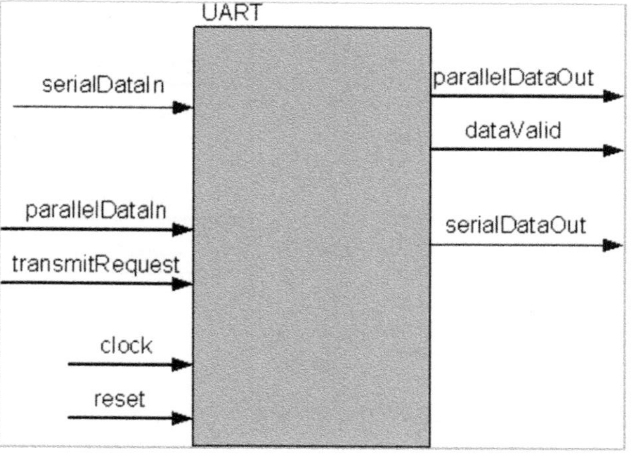

Figure 3.19: Block Diagram for the Top of the UART Design.

Where signals for the top level of the UART are summarized as follows:

Table 3.13

Signal Name	Type	Direction	Description
reset	std_logic	in	Resets the module
clock	std_logic	in	System clock
serialDataIn	std_logic	in	Arriving serial data
parallelDataOut	slv[1] (7 downto 0)	out	Successfully received character
dataValid	std_logic	out	Pulse indicating parallelDataOut has become valid
parallelDataIn	slv(7 downto 0)	in	Character to transmit
transmitRequest	std_logic	in	Command to begin transmission of parallelDataIn
serialDataOut	std_logic	out	Serialized version of parallelDataIn

[1]For the sake of table space, std_logic_vector is abbreviated slv.

While Figure 3.19 is useful for identifying which signals are input and which are outputs and the types of signals, we need to go "deeper".

We know that a UART has at least two parts: a transmitter and a receiver. The clocks provided to the typical FPGA are usually on the order of dozens if not hundreds of MHz. UARTs typically transmit at a rate of 300–115,200 (although rates both higher and lower are defined). This implies some sort of clock modification module.

Since this is a simple example, let's make this UART a single baud rate using 8-data bits and one stop bit with no parity.

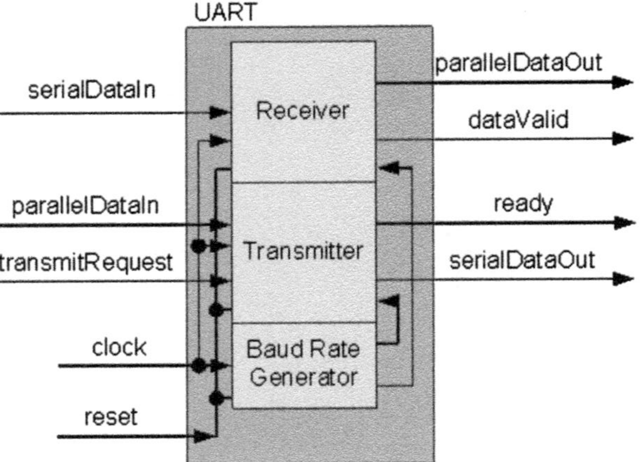

Figure 3.20: Peering down 1 hierarchical level into the UART.

Although at this point in time we may not have yet figured out how each of these three modules are constructed, we can figure out how they connect and produce the block diagram shown in Figure 3.20.

From Figure 3.20, let's identify all of the signals for each submodule and provide a brief description of how it may work...

3.1.8 Baud Rate Generator

The Baud Rate Generator module, when not in reset, divides the clock into two frequencies: that of the serial data rate for the outgoing serialDataOut signal, and the serial data rate x16. This higher rate is used to sample the serialDataIn (as this is an asynchronous signal). It is important to note that these frequencies derived from the clock are NOT used to clock the Transmitter module and Receiver module, but rather to enable them at the appropriate time.

Table 3.14

Signal Name	Type	Direction	Description
reset	std_logic	in	Resets the module
clock	std_logic	in	System clock
baudRateEnable	std_logic	out	Produces one clock-width pulse every *N* clocks
baudRateEnable_x16	std_logic	out	Produces one clock-width pulse every *N*/16 clocks (16 times faster than baudRateEnable)

Best Design Practices 3.8 – Tips

Never drive a clock input with a signal that was generated by FPGA fabric (LUTs or Flops)!
Corollary: clocks may only be divided by special silicon resources.
Corollary: Use clock enables to have logic run at a lower rate.

3.1.9 Transmitter

The transmitter behavior is very straightforward, parallelDataIn is registered when transmitRequest is asserted. In this fashion it is held static during the transmission of the bitstream. The transmitter will use the baud rate provided by the baud rate (enable) generator. A small-state machine can be used to shift the data out through serialDataOut.

Table 3.15

Signal Name	Type	Direction	Description
reset	std_logic	in	Resets the module
clock	std_logic	in	System clock
baudRateEnable	std_logic	in	Advance to the next bit to transmit when this signal is high
parallelDataIn	slv(7 downto 0)	in	Character to transmit
transmitRequest	std_logic	in	Command to begin transmission of parallelDataIn
ready	std_logic	out	Indicates that the transmitter isn't currently transmitting
serialDataOut	std_logic	out	Serialized version of parallelDataIn

3.1.10 Receiver

The Receiver is a bit more complex as it must automatically synchronize itself with the arriving data. The baud rate x16 enable will sample serialDataIn and look for a high-to-low transition which marks the beginning of the serial data packet. Once this point is found, the middle of each data bit is known and can be captured into a simple shift register. Once all the bits are received, the data can be presented on parallelDataOut and dataValid can be asserted.

Table 3.16

Signal Name	Type	Direction	Description
reset	std_logic	in	Resets the module
clock	std_logic	in	System clock
baudRateEnable_x16	std_logic		Sample *serialDataIn* each time this signal is active
serialDataIn	std_logic	in	Arriving serial data
parallelDataOut	slv (7 downto 0)	out	Successfully received character
dataValid	std_logic	out	Pulse indicating parallelDataOut has become valid

If we were to draw a hierarchical drawing of this module, it would look as shown in Figure 3.21.

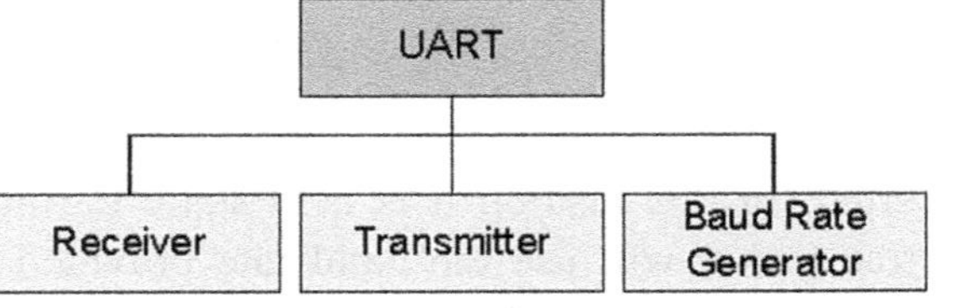

Figure 3.21: UART Hierarchy (so far...).

We'll break a cardinal rule for the moment – we should code up the three submodules and build simple test benches to prove that these modules are working as they should. The reason that we're breaking this rule here is that at this point in the book, we have enough information to build this higher-level structure using component and signal instantiation and combinatorial assignments, but we haven't yet come to a discussion of processes which will greatly ease the effort of developing these remaining modules. So, we'll take what we have for now and finish off this design later in the book when we discuss processes. For now, let's code up what we can. First we can build the entities for the three modules based on our tables...

```
entity UART_baudRateGenerator is
   port (reset               : in  STD_LOGIC;
         clock               : in  STD_LOGIC;
         baudRateEnable      : out STD_LOGIC;
         baudRateEnable_x16  : out STD_LOGIC
        );
end entity UART_baudRateGenerator;
```

Snippet 3.45: Source Code for the Entity Statement for UART_baudRateGenerator.

```
entity UART_transmitter is
    port ( reset           : in  STD_LOGIC;
           clock           : in  STD_LOGIC;
           baudRateEnable  : in  STD_LOGIC;
           parallelDataIn  : in  STD_LOGIC_VECTOR (7 downto 0);
           transmitRequest : in  STD_LOGIC;
           ready           : out STD_LOGIC;
           serialDataOut   : out STD_LOGIC
         );
end entity UART_transmitter;
```

Snippet 3.46: Source Code for the Entity Statement for UART_transmitter.

```
entity UART_receiver is
   port (reset              : in  STD_LOGIC;
         clock              : in  STD_LOGIC;
         baudRateEnable_x16 : in  STD_LOGIC;
         serialDataIn       : in  STD_LOGIC;
         parallelDataOut    : out STD_LOGIC_VECTOR (7 downto 0);
         dataValid          : out STD_LOGIC
        );
end entity UART_receiver;
```

Snippet 3.47: Source Code for the Entity Statement for UART_receiver.

Although these entities can be placed in a single file, this is rarely done. Typically each entity/architecture pair is placed in its own file and the VHDL/synthesis/implementation development environment manages these files for us. Usually the file name is just the name of the entity + ".vhd" or ".vhdl".

Best Design Practices 3.9–Tip

Place each entity and its associated architecture in its own file.

Referring back to Figure 3.21 we see how all three of these modules are connected, thus we should be able to construct the top level of the UART. First the entity:

```
entity UART is
    port (reset             : in  STD_LOGIC;
          clock             : in  STD_LOGIC;
          serialDataIn      : in  STD_LOGIC;
          parallelDataOut   : out STD_LOGIC_VECTOR (7 downto 0);
          dataValid         : out STD_LOGIC;
          parallelDataIn    : in  STD_LOGIC_VECTOR (7 downto 0);
          transmitRequest   : in  STD_LOGIC;
          txIsReady         : out STD_LOGIC;
          serialDataOut     : out STD_LOGIC
         );
end entity UART;
```

Snippet 3.48: UART Entity Statement.

Next, the three components: UART_transmitter, UART_receiver, and UART_baudRate Generator must be defined between the architecture and begin statements:

```
architecture Behavioral of UART is

 component UART_baudRateGenerator
    port(reset               : in  std_logic;
         clock               : in  std_logic;
         baudRateEnable      : out std_logic;
         baudRateEnable_x16  : out std_logic
        );
 end component UART_baudRateGenerator;

 component UART_transmitter
    port(reset            : in  std_logic;
         clock            : in  std_logic;
         baudRateEnable   : in  std_logic;
         parallelDataIn   : in  std_logic_vector(7 downto 0);
         transmitRequest  : in  std_logic;
         ready            : out std_logic;
         serialDataOut    : out std_logic
        );
  end component UART_transmitter;

  component UART_receiver
     port (reset              : in  std_logic;
           clock              : in  std_logic;
           baudRateEnable_x16 : in  std_logic;
           serialDataIn       : in  std_logic;
           parallelDataOut    : out std_logic_vector(7 downto 0);
           dataValid          : out std_logic
          );
  end component UART_receiver;
```

Snippet 3.49: Component Definitions for UART_receiver, UART_transmitter, and UART_baud RateGenerator.

Next, we'll define the signals that connect the module together. These signals are clearly seen in the detailed block-level illustration. Note that only *baudRateEnable and baudRateEnable_x16* are defined in this section. The other signals (*reset, clock, etc.*) are "defined" within the port section of the entity.

```
signal baudRateEnable      : std_logic := 'U';
signal baudRateEnable_x16  : std_logic := 'U';
```

Snippet 3.50

Finally, between the "begin" and "end" statements of the architecture, we can instantiate the three modules and "wire" them together (structural coding style):

```
begin

      xmit: UART_transmitter port map (
            reset            => reset,
            clock            => clock,
            baudRateEnable   => baudRateEnable,
            parallelDataIn   => parallelDataIn,
            transmitRequest  => transmitRequest,
            ready            => txIsReady,
            serialDataOut    => serialDataOut
      );

      rcvr: UART_receiver port map (
            reset              => reset,
            clock              => clock,
            baudRateEnable_x16 => baudRateEnable_x16,
            serialDataIn       => serialDataIn,
            parallelDataOut    => parallelDataOut,
            dataValid          => dataValid
      );

      rateGen: UART_baudRateGenerator port map (
            reset              => reset,
            clock              => clock,
            baudRateEnable     => baudRateEnable,
            baudRateEnable_x16 => baudRateEnable_x16
      );

end architecture Behavioral;
```

Snippet 3.51: Body of the UART's Architecture.

Often a vendor's tools will provide some mechanism to view the results of synthesis in a graphical format. Xilinx offers the "RTL Schematic" viewer which outputs the schematic shown in Figure 3.22 as a result of the above design.

This type of tool helps the designer visualize how the modules are hooked together. If there were additional modules below any of these blocks, the user could descend into the module to further confirm connections.

It is important to note that this diagram is based strictly on the VHDL code and does not necessarily represent how this design will be implemented in the FPGA. Once we build the

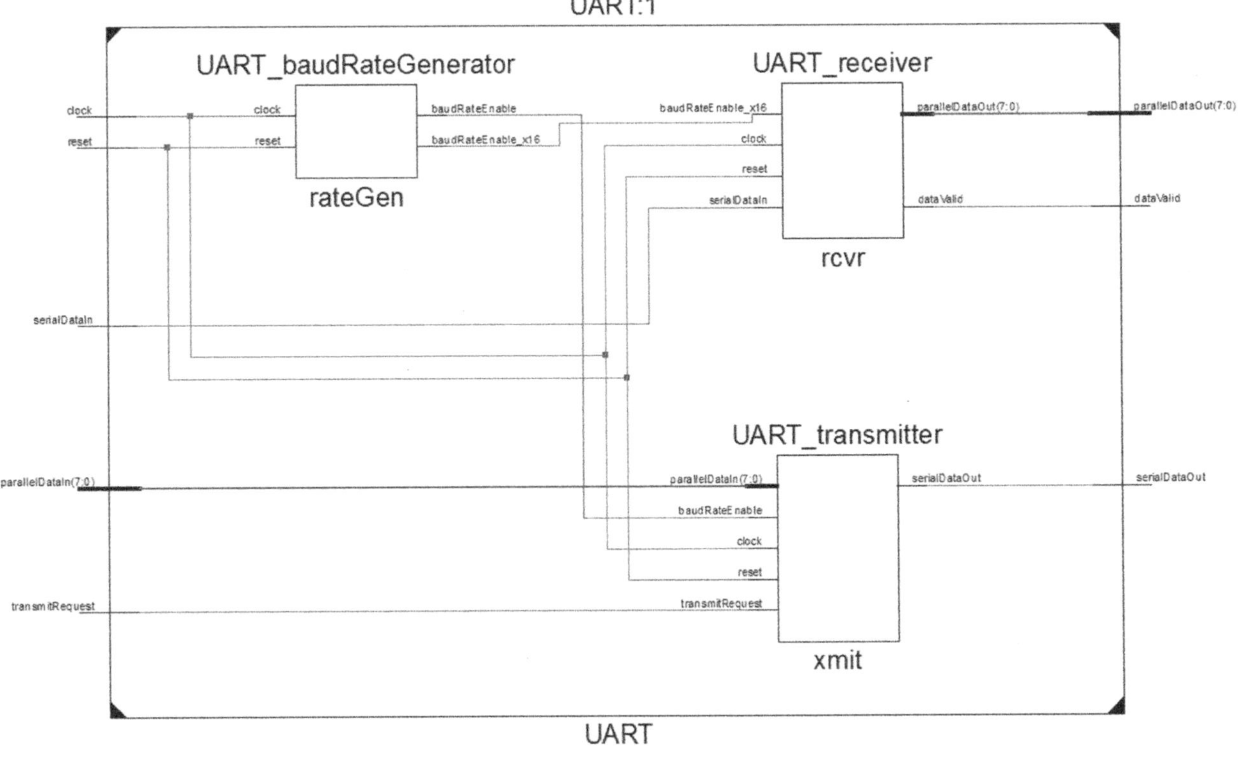

Figure 3.22: Schematic result from the RTL viewer.

lower-level contents for each module, we will revisit the RTL viewer and the Technology Viewer which will show us how the tools will implement the design in silicon.

SIDEBAR 3.2

Hardware vs. Software Programming

Good hardware and software design relies on hierarchy – the structure created from smaller blocks of a design.

Software terms these blocks "functions", "methods", "procedures", "subroutine", and the like depending on the language. When one of these modules is "called", the computer marks where the "jump" to this module occurs, usually by pushing a pointer to the program counter onto the stack, followed by some number of arguments being pushed onto the stack, and reserves space on the stack for whatever local memory is required by the function.

Since a CPU is a single-instruction-at-a-time device, all of these operations that must occur to call a subroutine cost time as well as run the risk of creating a run-time error due to the possibility of a stack overflow. This diminishes performance and reliability.

In software, this can be avoided by writing "in-line" functions, which effectively copies the entire subroutine each time it is used. This improves execution speed and avoids stack overflow, but requires significantly more program memory space.

VHDL, like software programming languages, has a same modular concept, but with some very significant differences – these "functions" are "unrolled" or turned into hardware during synthesis (as opposed to run-time) and affect how many resources are consumed. These modules ALWAYS exist regardless of how much data are pumped through (or even if they are ever enabled or not)! Similarly, circuits can be designed to "time-share" where the overall performance is diminished, but with the savings of some amount of logic.

So, in effect, hardware and software function implementations are opposites – in software it costs time and in hardware it costs space. There are coding styles in both hardware and software that allow the reverse to occur – in software "in-lining" takes up more space, but runs faster, in hardware "resource-sharing" costs performance, but requires fewer hardware resources.

3.1.11 Introducing the Simulation Environment

We've given a fair bit of lip service to simulation and particularly how some things in simulation differ from how they may be implemented in hardware.

Let's take a few minutes and discuss how simulation is performed and how you can use it to validate your hardware before moving to the bench.

Simulation generally requires one additional layer of hierarchy that synthesis does not. The top level of the design in synthesis becomes the interface with the world outside the FPGA in which connections to "virtual hardware" can be made (for example A/Ds and memory). Since simulation is a software abstraction, your code describes the FPGA, but no "outside world" – unless you create one.

This "outside world" which provides stimulus to the FPGA and displays the output produced by the FPGA is another VHDL file referred to as a "test bench". Depending on the level of sophistication that is needed to validate the Unit Under Test (UUT), which is the top level of your FPGA design, or some portion thereof, the test bench can be as simple as a clock source and simple stimulus to prove out some submodule, or incredibly complex, modeling a full system including other FPGAs, processors, memories, Analog-to-Digital converters, Digital-to-Analog converters, and models of real-world input such as temperature, velocity, voltage, etc.

Regardless of the level of complexity, the Simulation Environment is generally responsible for displaying the waveforms for the input (stimulus) and output (response), providing the user interface including which signals are visible, how signals are grouped, determining how the signals are displayed (color, radix, etc.), and possibly providing additional diagnostic tools such as code single stepping, breakpoints, and visualization tools.

Often designers overlook simulation during the development of the project, instead performing simulation only at the end of coding process. There are those that skip the simulation stage altogether, instead favoring testing in hardware. This is not good design practice! In fact simulating small pieces of code, verifying their functionality, then assembling the proven modules into larger constructs and simulating them is a much better practice!

Best Design Practices 3.10 – Tip

Simulate each module as you code from the bottom up so that you will have confidence that your lower level modules are working properly before you tackle the larger (i.e. more complex) higher-level modules.

Simulation can be performed at a number of points throughout the development process. The fastest, easiest, and most commonly performed simulation is known as the functional or behavioral simulation. Your VHDL code, even before synthesis, is analyzed and simulated. Although there isn't any timing information pertaining to where various resources are placed within the FPGA, this type of simulation can quickly "prove" that your code is behaving as expected.

Best Design Practices calls for us to design from the "top-down", and code from the "bottom-up". Therefore, we should always be coding the leaf-level (lowest) portion of the design first. These modules are usually fairly simple and can be tested with fairly simple stimulus. Once you are more familiar with VHDL and coding test benches, it often takes a few minutes to code up an appropriate test bench to verify that what you've coded is working the way that you expect it to. Armed with the knowledge that your lowest-level modules are behaving appropriately, you can then tackle the higher-level modules which are usually just a couple of lower-level modules stitched together with some additional logic.

Only when the higher levels of a large design are encountered does the generation of stimulus become a bit more challenging – such as writing an accurate simulation of Ethernet packets, Radar returns, or complex feed-back. This will require more advanced simulation techniques for which there are a number of solutions including capturing "live" data and playing back into simulation, or learning to write very detailed test benches for which there are numerous books written.

Simulation can also occur after the VHDL code has been synthesized (i.e., netlist simulation) which demonstrates that the synthesis tools accurately converted your VHDL code into a netlist (provided, of course, that there aren't differences between your behavioral simulation and your netlist simulation).

Recall that early on we discussed two aspects to an FPGA design – "does it do what it is supposed to do (behavior)?" and "does it do it fast enough (performance)?" The following simulations begin to take the performance aspect into account.

Further in the flow as the design makes the transition into hardware, there is post-Map simulation where some of the timing delays due to non-routing FPGA resources are known (i.e., flop, LUTs, BRAM, DSP elements, IOs, etc.), and finally post-Place-and-Route simulation in which all the delays in the FPGA are known.

Both post-Map and post-Place-and-Route simulations are generally not performed for two reasons: first, they are very time-consuming for anything larger than a small or medium design. Second, another tool – the Static Timing Analyzer – is available. This tool analyzes all of the paths through the FPGA against the timing constraints provided to the implementation tools and reports unconstrained paths (paths that are not covered by a timing constraint) and failing paths (paths that take too long to meet timing).

Resolving timing issues is a topic for other books as it has more to do with the specific set of implementation tools than the language. That said, poor coding techniques can lead to an inability to close timing.

The unit/device under test

The UUT or DUT represents the top level of the design that you intend to synthesize. Typically, this level doesn't perform much logic, but rather stitches together many subordinate hierarchical modules. A typical UUT/DUT module usually consists of input and output buffer instantiations, instantiation of clock resources, and instantiations of the actual design modules and any other modules required to support the modeling of the device or system (see Figure 3.23).

One of the highly beneficial aspects of simulation is that more than just the signals that are present at the top level of the design are available for inspection. Generally, any signal found in the design can be viewed.

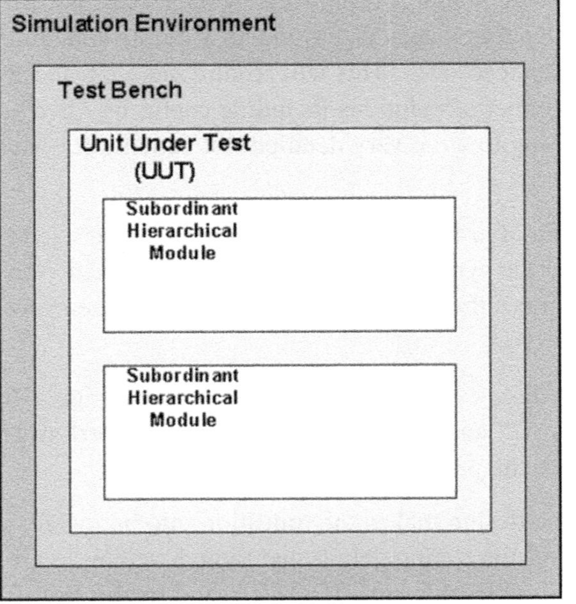

Figure 3.23: Hierarchical Diagram of the Simulation Environment.

Continuing our example of an RS-232 UART…

Recall that the top level of the UART design has the signals that are shown in Figure 3.24.

We can construct a test bench which instantiates the UART and provides a clock source, reset source, and stimulus to get the device going (see Figure 3.25). We can even be a little tricky and feed the serialDataOut back into the serialDataIn and ensure that the parallelDataOut matches the parallelDataIn (albeit delayed by the serialization/deserialization).

Each of these blocks in the test bench can be implemented as a line of code, a process (which we will see in the next section), or an instantiated module.

The Reset Generator is tasked with producing an active high signal for some period of time after power is applied (this is equivalent to "once the simulation has begun"), let's arbitrarily set this value to 100 ns. After the 100 ns has passed, reset should de-assert. Borrowing from an earlier code snippet:

```
reset <= '1', '0' after 100 ns;
```

Snippet 3.52

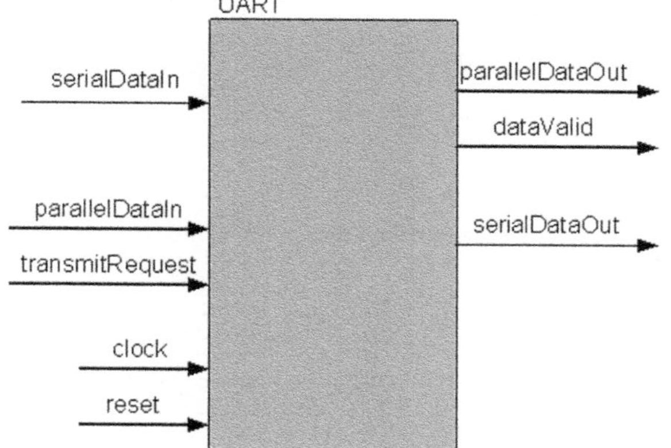

Figure 3.24: Recalling the top level of the RS-232 UART.

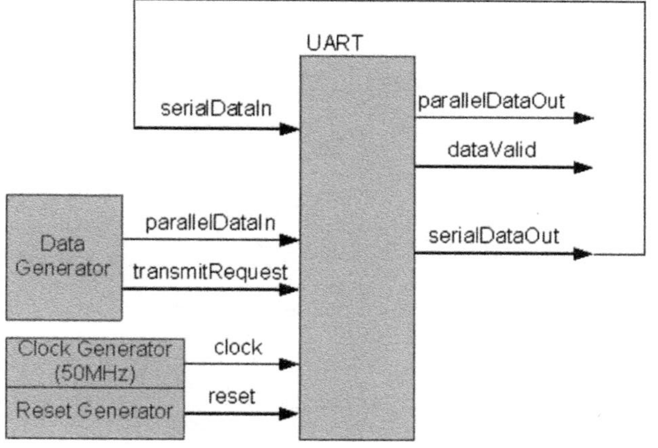

Figure 3.25: Additional logic to form a test bench.

The clock generator we saw illustrated earlier for generating a 50% duty cycle clock at 100 MHz.

```
constant clock_period : time := 10 ns;      -- 100MHz
signal   clk          : std_logic := '0';   -- '1' or '0'
. . .
clk <= not clk after clock_period / 2;
```

Snippet 3.53

The data generator module is most easily coded using a process statement which we're coming to shortly.

So far, our test bench code for the UART looks like this…

```
LIBRARY ieee;
USE ieee.std_logic_1164.ALL;
USE ieee.numeric_std.ALL;

ENTITY UART_design_tb IS
END UART_design_tb;

ARCHITECTURE behavior OF UART_design_tb IS

  -- Component Declaration for the Unit Under Test (UUT)
  component UART
    port (
      reset            : IN  std_logic;
      clock            : IN  std_logic;
      serialDataIn     : IN  std_logic;
      parallelDataOut  : OUT std_logic_vector(7 downto 0);
      dataValid        : OUT std_logic;
      parallelDataIn   : IN  std_logic_vector(7 downto 0);
      transmitRequest  : IN  std_logic;
      serialDataOut    : OUT std_logic
     );
  end component UART;

  --Inputs
  signal reset : std_logic := 'U';
  signal clock : std_logic := '0';
  signal serialDataIn      : std_logic := '0';
  signal parallelDataIn : std_logic_vector(7 downto 0) :=
                                  (others => 'U');
  signal transmitRequest   : std_logic := 'U';

  --Outputs
  signal parallelDataOut : std_logic_vector(7 downto 0);
  signal dataValid       : std_logic;
  signal serialDataOut   : std_logic;

  -- Clock period definitions
  constant clock_period  : time := 10 ns;

begin

  -- Clock and reset creation
  clock <= not clock after clock_period / 2;
  reset <= '1', '0' after 100 ns;

  -- Instantiate the Unit Under Test (UUT)
  uut: UART PORT MAP (
         reset           => reset,
         clock           => clock,
         serialDataIn    => serialDataIn,
         parallelDataOut => parallelDataOut,
         dataValid       => dataValid,
         parallelDataIn  => parallelDataIn,
         transmitRequest => transmitRequest,
         serialDataOut   => serialDataOut
       );

end architecture behavior;
```

Snippet 3.54: Test Bench for UART (incomplete).

Lines 1–3: Access to the IEEE library indicating the anticipated use of the std_logic package and its associated numeric_std package.

Lines 5–6: Notice that this entity has no port information. This looks odd, but consider: the top level of the UUT contains the description of the inputs and outputs of the FPGA which connect to the outside world. Where does the test bench connect to? The answer is *nowhere*. The test bench exists only in simulation and all signals are generated and monitored in this environment.

Line 8: beginning of the architecture for this entity.

Lines 10–22: component definition for the UUT which is the UART module.

Lines 25–29: signal definitions for the signals which connect to the inputs of the UUT. Notice the presence of initializations. Why are some signals set to 'U' while others are zeroed? The clock signal is set to zero because the code relies on a known starting value so that it can oscillate. As this code is written, the clock will begin oscillating at a logic high. If you wanted the clock to begin at a logic log, then line 26 can be changed from a '0' to a '1'. Other signals are left as 'U' so that any failure to properly initialize these signals in the code is clearly seen and can be fixed.

Lines 32–34: signal definitions for the output of the UUT.

Line 37: constant definition for the clock_period. The time that this is set to indicates the duration for a single period of the clock. The code divides this value in half to produce a 50% duty cycle. The frequency of the clock is easily changed by modifying the time indicated in this line.

Line 39: beginning of the behavioral description for this architecture.

Lines 41–51: instantiation of the UUT. Certainly the signals that are assigned to this module do not need to be identical to the port name; however, unless the port names are horribly cryptic, this is typical as these are the signal names that will appear in the simulation.

Line 54: generates a 50% duty cycle at the specified clock frequency.

Line 55: generates an active reset for the first 100 ns after simulation begins.

Line 57: conclusion of the behavioral description of the architecture.

After we discuss processes, we will code the lower-level modules and return to this test bench and simulation of the full UART design.

Finally, we can make the final connection which loops the *serialDataOut* output from the transmitter portion of the UART back into the *serialDataIn* input to the receiver portion of the UART. This is easily done with one assignment: *serialDataIn <= serialDataOut;*.

The question becomes, where to put this line of code? Before the instantiation of the UART module, after it, but before the clock and reset generation, someplace else?

From VHDL's point of view, it just doesn't matter! VHDL considers everything between the architecture's *begin* statement and *end* statement to occur simultaneously. From a human point of view many designers choose to order their modules and concurrent statements in a similar fashion as schematics and flow diagrams are drawn. Graphic representations of the design tend to show information flowing from the left side of the page to the right side and textual description of the design from the top to the bottom.

The astute reader will notice that there is nothing providing the input for *serialDataIn*. We'll cover this process at the end of the next iteration.

CHAPTER 4

Loop 2 – Going Deeper

The previous Loop introduced the Entity and Architecture structures in which the description of how our design connects to the outside world and how our design actually works is contained. This loop also introduced us to signals and what can be done to signals (in their concurrent form) and the concept of a test bench.

This Loop introduces the Process statement in which we can encapsulate sequential statements and use the variable construct (which behaves somewhat differently than the signal) and some of the myriad things that can be done in the "sequential" realm.

4.1 Introducing Processes, Variables, and Sequential Statements

Remember that concurrent statements are all "executed" simultaneously. This works extremely well when modeling simple gates and flops and even black boxes. So, can't everything be modeled this way? Sure! But it can be a real pain in the backside! Sometimes it is easier to represent a design as a sequence (i.e., do this first, then this, then this...). We'll look at that here.

VHDL provides such a mechanism in the form of the "Process". In short, a Process encapsulates a collection of variables and statements that is executed in the order in which the statements are listed.

Processes look like "mini-architectures" in their general form. The process begins with the "process" keyword and the name of the process. This is followed closely by a list of signals which the process is "sensitive" to. This line is roughly analogous to an entity statement (there are a few significant differences, most notably that all items defined in the architecture in which this process resides are still available within the process – the "gateway" that the entity provides is absent here). Next is a begin statement followed by sequential statements and ended with the end clause.

Before we can effectively discuss stuff that can get shoved into a process, we should discuss two aspects which are unique to processes. First is the concept of the "sensitivity list" and the other is that of "process suspension".

Recall that one of VHDL's primary reasons for existence was simulation and that the language was constructed a time long ago when computer cycles were very expensive. This was one of the motivating reasons for the "process sensitivity list". In short, this list was an

VHDL 101.

enumeration of all the signals inside the process that *could cause* a change. Let's look at the following example of an "AND" gate:

```
process and_gate (a,b)
  begin
   c <= a and b;
  end process and_gate;
```

Snippet 4.1: Example of Asynchronous Process.

If either *a* or *b* change then *c* would need to be "recomputed" (although it might not actually change value). When implemented in hardware, this simply becomes an "AND" gate with the '*a*' signal driving one input and the '*b*' signal driving the other. The output, '*c*', always reflects the (nearly) instantaneous changes on *a* or *b*. Therefore, there is no need to "check" for changes on the input since it is implicitly covered in the implementation of the gate within the silicon.

Simulation, however, is another story. As the simulator reaches this process, it will look at the process sensitivity list, and, in order to avoid simulating this process unnecessarily and wasting expensive clock cycles, the tools would check to see if either *a* or *b* changed, and if either or both have, this process would be "run" (simulated) otherwise it would be skipped. While it is true that for such a short process it might take as much time to see if this process should be run rather than just running it, it does make sense since most processes are considerably more complex, and hence, time-consuming.

What happens if a signal is accidentally, or even deliberately, excluded from the process sensitivity list?

Well, it depends whether this design is being synthesized (targeting hardware) or simulated!

The process sensitivity list is essentially ignored when you synthesize the design (i.e., generate a netlist which will be converted into a bitstream for your FPGA). So the answer in this context is "nothing". On the other hand, if you are running a simulation (which you should almost always do!) your compiler may issue a "simulation mismatch" and provide a warning indicating which signal you left out of the process sensitivity list. A "simulation mismatch" warning is an indication that your simulation may not run like the actual implementation – a big problem to be avoided as that is why we do simulation in the first place!

The obvious question becomes, "well, if the tools are smart enough to detect that a signal is missing from the list, why doesn't it just add it?" One answer is that VHDL is a very formal language and that everything needs to be declared explicitly so that there is no question that the code generates what the user describes (as opposed to what the user may want). This represents a full and complete description – if the tools were to change the meaning (even if it's something that you wanted), then it would no longer be a complete meaning as the code would then be open to interpretation by the tools (which, by their very nature vary) and hence ambiguity would be introduced into a process that is supposed to remove ambiguity.

SIDEBAR 4.1

Synchronous Design Tip

Tip: Here's a quick way to minimize problems with the process sensitivity list: make (almost) all of your processes synchronous!

This is a good design practice as FPGA are loaded with flip-flops. The way that today's architectures are constructed, combinatorial logic is computed in an asynchronous look-up-table (LUT) and is paired with a flip-flop (go back to the introduction to FPGA architecture (Figure 1.2: Generic Slice). If the LUT is used and bypasses its paired flip-flop as is the case in a purely combinatorial design, then the flip-flop goes unused for this design. Similarly, if the flip-flop is used, and the LUT circumvented, then the LUT cannot be used for anything else either. So, if you use a LUT (combinatorial logic) you might as well use the associated flip-flop.

Good synchronous design practices will reduce the time spent debugging your design and help you meet your performance (timing) goals.

One quick note: if each two-input piece of combinatorial logic is latched, then there are two problems that arise:

1. The 2 input combinatorial logic can't be combined to make efficient use of the 4 (or 6) input LUTs leading to consumption of far more resources than necessary (which also translates to wasted power and extra heat that must be removed), and
2. Huge clock latencies. Even though your design could run at nearly the top performance of the FPGA, there will be so many unnecessary pipelined logic modules that there may be significant latency before the solution can be realized. This leads to one more problem – balancing of internal delays so that the right answers appear at the inputs to other combinatorial logic at the right time.

As a rule-of-thumb – keep your processes to a single thought. It makes it easier to code, easier to debug, and breaks the design into pieces that are neither too large nor too small.

By making your process purely synchronous, you will only have one signal in your process sensitivity list – the clock!

Let's look at what is happening...

```
clocked_AND_gate: process (clk)
   begin
     if rising_edge(clk) then
        c <= a AND b;
      end if;
   end process clocked AND gate;
```

Snippet 4.2: Clocked AND Gate.

1. The signal "clk" is included in the process sensitivity list. That means that every time the clock changes state (either rises OR falls), this process will be evaluated.
2. Line 3 uses the IEEE function "rising_edge()" to test if the clock's edge is rising or falling. If the edge is rising, then the true clause of the if statement is executed, otherwise nothing happens. Note that the true clause will be executed 50% of the time as this process is evaluated on BOTH edges of the clock, but the if statement stipulates that action occurs on only one of those edges.
3. The "true clause" of the "if" statement is our old friend the "AND" gate.

Here is the equivalent schematic of this process...

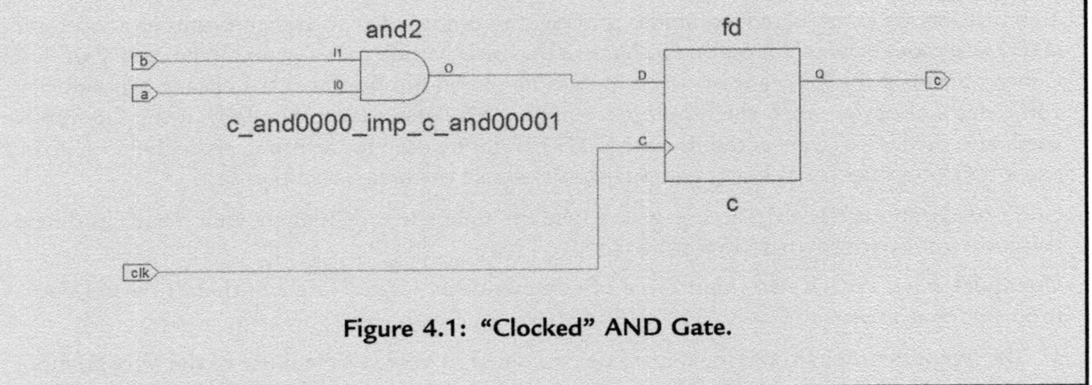

Figure 4.1: "Clocked" AND Gate.

How does this process structure get handled in both simulation and synthesis?

Understanding the hardware behavior is easy – during the synthesis process all the statements in the process are converted to a static set of silicon resources: LUTs, Flip-Flops, shift registers, block memory, etc., and connected together appropriately (recall that the specific placement within the FPGA of these items doesn't occur until the last stages of implementation). Basically, it becomes "always running" hardware.

Simulation is quite a bit different.

At first blush, you might say that the instructions are executed in a sequential fashion from the begin statement of the process to the end statement. You'd be correct. But what happens then? The end statement acts like a "goto" statement and jumps to the beginning of the process (not the begin statement). In this fashion, a continuously executing loop of code is created.

But how does the simulation tool create a difference between continuously “running” hardware and getting stuck in an infinite loop?

Simulation introduces the concept of “process suspension”. Processes enter suspension when same form of the “wait” statement is encountered.

Since we’ve already discussed the process sensitivity list, we might guess that when the process sensitivity list is encountered, the simulation waits there (suspends) until a change in one of the members of the process sensitivity list occurs. This is one mechanism for preventing the dreaded infinite loop.

The general form of the wait statement is as follows:

```
[ <label> : ] wait [ <sensitivity_clause> ]
                   [ <condition_clause> ]
                   [ <timeout_clause> ] ;
```

Snippet 4.3

Note the order of the optional portions of this statement. While it is possible to use all of the optional parts of the wait statement, typically none or only one of these optional clauses is used at a time giving us the apparent wait forms of:

- wait
- wait for <time>
- wait until <event>
- wait on <signal list>

While it is true that the process sensitivity list is a form of a wait statement, it is tolerated during synthesis, but it has no effect on the code. That is, the synthesis tools don’t complain about it because it is necessary for simulation, it just doesn’t do anything with it!

The following explanations are valid for simulation only and will generally cause errors if synthesized. Certainly any wait statement that uses time is not going to be synthesizable as time itself is not a synthesizable quantity. Other forms of the wait statement may be synthesizable based on the specific support provided by the synthesis tools.

“Wait” is the simplest form of this statement (i.e., no optional parts). When “wait” is encountered during a simulation, the process stops operating (regardless of the contents of the process sensitivity list) – period. Other processes may continue to operate, but not one in which the “wait”, has been reached. This is most frequently used to stop a specific process without stopping the entire simulation.

"Wait for <time>" is typically used to suspend the process for a specific amount of time. Snippet 4.4 demonstrates the use of "wait for <time>" to construct a 100-MHz (10 ns) clock with a 60/40 duty cycle.

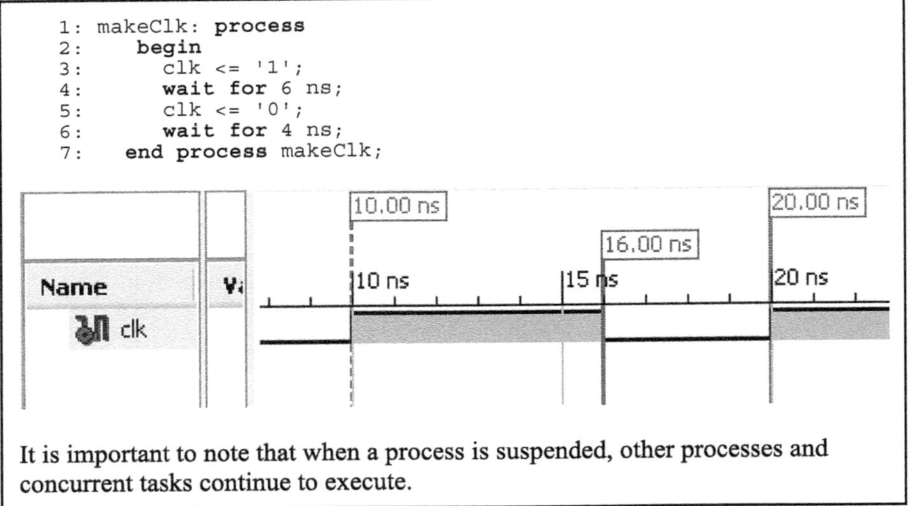

```
1: makeClk: process
2:     begin
3:       clk <= '1';
4:       wait for 6 ns;
5:       clk <= '0';
6:       wait for 4 ns;
7:   end process makeClk;
```


It is important to note that when a process is suspended, other processes and concurrent tasks continue to execute.

Snippet 4.4: Common Construct for Modeling a Clock.

Code analysis:

Line 1: opening to process named makeClk. Notice the absence of a process sensitivity list. This is an indicator to the simulation tools to always run this process.

Line 3: clk is driven high instantly.

Line 4: the process suspends for 6 ns effectively "holding" the clk signal high for this amount of time.

Line 5: clk is driven low instantly.

Line 6: the process suspends for 4 ns effectively "holding" the clk signal low for this amount of time.

Line 7: process statements end and implicitly jumps back to line 1.

The wait statement can be used to introduce jitter into a clock signal. There are Institute of Electrical and Electronics Engineers (IEEE) packages which contain math functions that can produce a random value which can be easily converted into time. If this "time" type variable is used as the argument to a "wait for" statement then the simulation will wait for that amount of time, even though that specific value changes each time the statement is reached.

```
 1: clkMan: process
 2:        variable halfCycleJjitter  : time;
 3:        variable rndVal0to1        : real;
 4:        variable seed1             : integer := 5;
 5:        variable seed2             : integer := 32;
 6:    begin
 7:      -- first the low portion of the clock
 8:      uniform(seed1,seed2, rndVal0to1);
 9:      halfCycleJitter:= (rndVal0to1- 0.5) * 50 ps;
10:      clock <= '0';
11:      wait for clock_period/2 + halfCycleJitter;
12:
13:      -- and now the high portion
14:      uniform(seed1,seed2, rndVal0to1);
15:      halfCycleJitter:= (rndVal0to1 - 0.5) * 50 ps;
16:      clock <= '1';
17:      wait for clock_period/2 + halfCycleJitter;
18:    end process clkMan;
```

Snippet 4.5

Code analysis:

Line 1: declaration of the process named *clkMan* notice the absence of a process sensitivity list – this is because (1) this process is not sensitive to anything happening externally to this process and (2) this process contains two wait statements which will cause the process to suspend and thus not enter an infinite loop.

Lines 2–5: variable declaration area.

Line 6: beginning of the description of the behavior of the process.

Line 8: uniform is a procedure in the IEEE.math_real package and produces a real value (*rndVal0to1*) between 0 and 1 based upon two integer seeds.

Line 9: *halfCycleJitter* offsets the random value by −0.5 effectively changing the range of *rndValue0to1* from 0 to 1 to −0.5 to 0.5. This value is then scaled by 50 ps which, for this example, illustrates the worst case jitter for this clock.

Line 10: clock is driven low.

Line 11: this process suspends for half of a clock period plus the jitter amount.

Line 14: another random value between 0 and 1 is computed so that the time during the low portion of the cycle is different than the time during the high portion of the clock cycle.

Line 15: this new random value is converted to a scaled time appropriate for this simulation.

Line 16: the clock is driven high.

Line 17: the process suspends for half of a clock period plus the new jitter amount.

Line 18: the end of the process causes the control of the process to jump to the process statement which then falls through and executes lines 8 through 10 where it then, once again, suspends on line 11. This process repeats until the simulation is terminated.

"Wait until <condition>" is used when the exact amount of time to pause is unknown. This is typically used to respond to a set of conditions that are compared using Boolean operators.

For example if we needed to wait until the rising edge of the receiver clock (clk_rx) so that we could either count the number of clock edges or to synchronize events, we could use the following statement:

```
1: wait until rising_edge(clk_rx);
```

Snippet 4.6: Waiting for a Specific Event.

The "until" clause can be coupled with the "for" clause as follows:

```
1: wait until rising_edge(clock) for 1 us;
```

Snippet 4.7

Code analysis:

Line 1: this wait statement can be read as "wait until a rising edge is seen on the clock signal, then wait for 1 μs".

Unfortunately, this order cannot be reversed to read "wait for 1 μs, then wait for the next rising edge". This would require the following two wait statements:

```
1: wait for 1 us;
2: wait until rising_edge(clock);
```

Snippet 4.8

Code analysis:

Line 1: wait for 1 μs regardless of what is happening on clock.
Line 2: once 1 μs has elapsed, wait for the next rising edge of clock.

Continuing on to the "Wait on" structure:

"Wait on <signal list>" is another structure used when the exact amount of time to pause for is unknown. This is typically used to respond to any change on any of the signals listed in the <event clause> events. The process sensitivity list is a special form of this type of wait statement since it is so commonly used to model combinatorial logic.

```
1a: process and_gate(a,b)
2a:     begin
3a:         c <= a and b;
4a:     end process and_gate;
```

```
1b: process and_gate
2b:     begin
3b:         c <= a and b;
4b:           wait on a,b;
5b:     end process and_gate;
```

Snippet 4.9: Equivalent "Waits" for AND Gate Processes.

Line 1a: the implied wait statement is built into the process sensitivity list. This line reads "keep this process in suspension until there is a change to either or both signals *a* or *b*.

This is equivalent to Line 4b.

Rule: every process must have some form of a wait statement in it.

Maxim 4.1

The bigger picture…

All this talk about processes and wait statement and we've almost skipped the real design questions – when/why do you use a process, and how big should it be?

Processes are used when it is more convenient to describe a behavior in a sequential fashion than a concurrent one.

Certainly the description of an AND gate, even a latched AND gate can be handled in one simple concurrent statement (Register Transfer Level (RTL) coding style):

C <= A and B when rising_edge(clk);

But do you really want to describe a state machine this way[1]?

How big (how many lines of code) should it be is a tougher question. The children's fairytale story of the Three Bears is silly, yet apropos. In the story, a little girl wanders into the home of an anthropomorphized family of three bears. Every experience she has in the house is grouped in threes – the items belonging to "papa bear", "mama bear", and "baby bear" – from eating the porridge ("too hot", "too mushy", "just right"), to the chairs ("too big", "too hard", "just

[1] It is certainly possible to! We've all done it in our junior level introduction to digital circuits classes. Was it fun? Easy? Efficient? Maintainable? Nope. We'll cover state machines shortly and see how much easier they are as processes!

right"), etc. She finally finds the bedroom and takes a nap on her selected bed ("too lumpy", "too wide", "just right") where she is found by the bear family and arrested for trespassing, theft, and vandalism.[2]

The real moral of this story is that the size of a process, like a module, like the design itself, should be "just right". This means that it should do only what it is supposed to, no more and no less. Simply put, it has to be "big enough" to get the job done and no more.

Sizing a module or process is somewhat experiential. For those who survived college in the 1980s, software subroutines were supposed to be no more than a page in length, including comments (which left about 20 lines of "real" code). The idea was to curtail "function creep"[3] (which could have been better handled by emphasizing a specification driven design flow) and keep the code modular for later reuse (which didn't happen as there still isn't a good code snippet repository tool!), and most importantly, to keep the designer focused on the task at hand.

If this discussion sounds like the same discussion we just had regarding hierarchy, it should! A design should be broken into large conceptual pieces. Each piece should be broken down into smaller function groups, and these function groups further decomposed (possibly through several more layers of hierarchy). Finally, when the decomposition into separate modules is complete, and the work is actually going to be done, then introduce the process. The process represents the last node in the design process.[4]

Best Design Practices 4.1 – How Big Should a Module Be?

A module should be only one "thought" in size.
Corollary – Any sufficiently large design should be decomposed into smaller thoughts.
Each thought should be a separate module, whether it can be implemented as a component, process, or RTL statement.
The lowest level of thought should be the action.
This implies that any action is the result of a well thought out process. Now if we could only apply this to our everyday lives, we'd really have something!

---

2 There are a number of endings to this story including the traditional European version in which she is eaten by the bears, the "New World" soft-and-kind version where she is simply scared and manages to outrun three outraged bears. Personally, I think the home was equipped with a burglar alarm and the police were there inside of 5 min.

3 Function creep is the natural tendency for an engineer or programmer to make the code more "capable" than the original intent. This unfortunate process leads to buggy code and delays in release.

4 This should not be construed to mean that processes can only reside in leaf (bottom) modules along any hierarchical path, rather, if there is a place for a process to be inserted to complete a thought, then it should be placed there. Many good designs see processes' places in the same module as component instantiations.

4.1.1 Variables

Variables are temporary storage mechanisms which reside within the process and whose scope is limited to just that process.[5] Variables behave somewhat differently than signals do – they take their assignment “instantaneously” – we’ll see this more clearly when we contrast the signal with the variable.

Variable declaration

Like signals, variables must be declared before they can be used. Since variables are specific to processes, they must be declared between the process clause and the “begin” statement of the process.[6] This also means that a variable is only valid within the process in which it is declared. Conveniently, variables follow the same naming and assignment rules that signals do.

```
1: varTest: process (x, y, z)
2:      variable i_var : integer range 0 to 100 := 3;
3:      variable r_var : real range 1.1 to 1.9 := 1.5;
4:      variable 3_choices : std_logic_vector(2 downto 0) := (others=>'0');
5:    begin
6:       variable internal : std_logic :=  'U' ;
7:    end process varTest;
```

Snippet 4.10: Examples of Variable Declaration – Both Legal and Not.

Code analysis:

Line 2: legal. Defines i_var to be of type integer and assume any value between 0 and 100. Sets initial value to 3.

Line 3: legal in simulation, illegal for synthesis. Real types are currently not supported by synthesis tools.

Line 4: illegal. Variables follow the same naming conventions as signals and may not begin with a number.

Line 6: illegal. All definitions must occur between the process statement and the begin. Behavioral descriptions occur between the “begin” and “end” statements.

Variable use

Variables take on the value that is assigned to them immediately. This is convenient as variables are often used to produce and test intermediate results. To the reader with software experience this will seem intuitive.

---

5 There is a construct known as a “shared variable” which allows a variable to extend its scope outside of a single process, however, this is not frequently used and may produce undesirable “side-effects” and is left for a more advanced book on the topic.

6 Variables are also used in functions and processes and this will be covered in a later section.

VHDL uses the ":=" assignment operator to help differentiate between a variable assignment and a signal assignment (which uses the "<=" symbol) and a conditional test for equality (just a single "=").

The most common use of variables is to hold intermediate values. Let's begin by recoding the earlier example of conditional assignment shown in Snippet 3.31 with how the same thing might be coded in a process. We'll start with the declaration of four quadrants as follows:

```
 1: whichQuadrant: process (angle)
 2:      variable Q1          : boolean := false;
 3:      variable Q2, Q3, Q4  : boolean := false;
 4:    begin
 5:      Q1  := (angle >=    0) and (angle <=  90);
 6:      Q2  := (angle >=   90) and (angle <= 180);
 7:      Q3  := (angle >=  180) and (angle <= 270);
 8:      Q4  := (angle >=  270) and (angle <= 360);
 9:      . . .
10:    end process whichQuadrant;
```

Snippet 4.11

Line 1: the beginning of a process which is named *whichQuadrant* and has a process sensitivity list of one member named *angle.*

Line 2: a variable named *Q1* is defined as a Boolean and initialized to false.

Line 3: three more variables, named *Q1, Q2,* and *Q3* are defined as Booleans and each is initialized to false.

Line 4: the begin statement separates the declarative portion of the process with the behavioral portion.

Line 5: *Q1* is assigned the value (note the use of the ":=") of true if the integer value of angle is equal to or greater than 0, but less than or equal to 90. If it is not, then *Q1* is assigned the value of false.

Lines 6–9: *Q2–Q4* are assigned values of true or false based on the angle.

Line 10: concludes the named process *whichQuadrant.*

Note that the process doesn't "do" anything. As it currently stands, it will be optimized away by the synthesis tools as it does not drive any signals. Also note that it may be possible for two values of the quadrant to be true simultaneously. This ambiguity will be resolved later using two mechanisms analogous to the conditional assignment with/select (the case statement) and the when/else (the if/then/else statement) structures.

Ordering of statements

Outside of a process, combinatorial assignments may occur in any order as they are "executed" simultaneously. Since statements within a process are "executed" sequentially, one must be careful to make certain of the ordering of the statements. Statements are executed from the processes' "begin" statement toward the processes' "end" statement. When the processes' end statement is reached, control jumps to the beginning of the process, and if there is a process sensitivity list present, the process suspends until any member of the process sensitivity list changes. Flow resumes from top to bottom and repeats.

Variables DO NOT reset to 0 or any initial value when the process suspends – they retain their value. Consider the following example:

```
1: cntMangr: process (someSignal)
2:       variable count : integer range 0 to 100 := 0;
3:   begin
4:      count := count + 1;
5:      . . .
6:    end process cntMangr;
```

Snippet 4.12

Code analysis:

Line 1: a process named *cntMangr* is defined with a process sensitivity list containing only *someSignal.*

Line 2: a variable named *count* is defined to be an integer between 0 and 100 and is initialized on power-up to 0.

Line 3: marks the end of the declarative portion of the process.

Line 4: count is incremented by one.

Line 6: end of the process.

Following the flow through the process we note that the process is not "executed" until there is a change on someSignal. If some signal is of type std_logic, then moving from a '1' to a '0' or a '0' to a '1' would both constitute a change and enable one "loop" of the process to be run. Similarly, a change from a '0' to a 'Z' would also constitute a change. Note that synthesis tools generally can't detect such a change, but the simulation tools can.

If someSignal were tied to a clock, then every transition that the clock made (both rising and falling edge) would cause count to increment.

Now a warning! If you are ever in a position where you must construct a purely combinatorial process, you must guarantee that the order of assignments flows from "top to bottom" and does not "wrap around" or require the use of the variable from the last pass through the process.

Here is an example of how NOT to code:

```
racing: process (a, b)
      variable v1, v2 : std_logic := 1;
   begin
      v1 := v2 nand a;
      v2 := v1 and (a or b);
      z  <= v2;
   end process racing;
```

Snippet 4.13: Creating Race Conditions.

Code analysis:

Line 1: a process named *racing* is defined to have a process sensitivity list of signals *a* and *b* which are defined as std_logic.

Line 2: two internal variables are defined (*v1, v2*) to be of type std_logic and initialized to '1'.

Line 3: end of the declarative portion of the process.

Line 4: *v1* is assigned the value of *v2* logically NANDed with *a*. Here's the question: what is the current value of *v2*? This is a bit tricky because this implies some kind of ability to "remember" the value of *v2* from line 5 (which hasn't yet been executed).

Line 5: *v2* is assigned the value of *v1* ANDed with the logical ORing of *a* and *b*. Same question as before: what is the value of *v1*? This time the answer is easy – *v1* was just computed on line 4.

Line 6: the signal *z* is assigned the value of *v2* which is now known since line 5 was "executed".

Xilinx's XST tool thinks the result shown in Figure 4.2 will be generated.

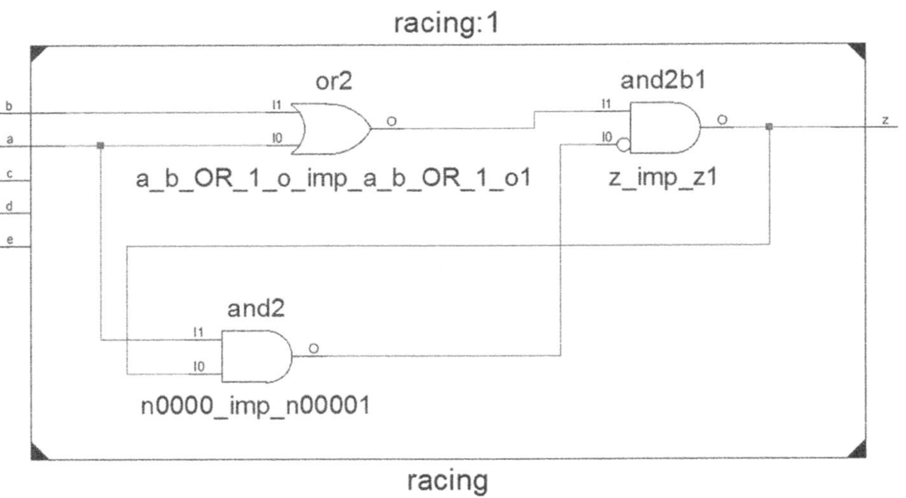

Figure 4.2: Result of Inappropriate Ordering of Variables.

Note the feedback signal from *z* back to the AND2 gate. What a nice way to build an oscillator![7]

[7] This is an incredibly sarcastic statement. In truth, this is a terrible thing to build! The device will oscillate at a frequency of roughly 1/(time through the LUT + feedback path). Not only will this mess up timing analysis, but if ever implemented in hardware would draw a considerable amount of power and cause unnecessary localized heating. There are those of you out there who think that this might be a clever way to build a clock circuit internal to the FPGA; however, it is not. Please don't build this! Use only the dedicated silicon resources inside the FPGA for clock generation and modification.

4.1.2 Signals within Processes

Signals within processes behave differently than they do when they are outside a process.

Outside the processes signals are essentially wire connections. Apply a voltage at one end, and the other end will have that same voltage[8] a small time later.

Inside the process a signal delays taking on its assigned value until the *process suspends.* So, as you can see from the example below, multiple assignments are made to a signal throughout the process, but the only value that "sticks" is the assignment that the signal had just prior to the process suspending.

```
1: mltplAsgns: process (a, b, c)
2:     begin
3:         Z <= a;
4:         Z <= b;
5:         Z <= c;
6:     end process mltplAsgns;
```

Snippet 4.14: Multiple Signal Assignments in a Process.

Code analysis:

Line 1: process named *mltplAsgns* is defined with a process sensitivity list of three (assume std_logic) signals named *a, b,* and *c*.

Line 2: end of process declarative area.

Line 3: signal *Z* (assumed to be defined as a std_logic) is assigned the value of signal *a*.

Line 4: signal *Z* is now assigned the value of signal *b*.

Line 5: signal *Z* is now assigned the value of signal *c*.

Line 6: end of the process, loops to line 1 and suspends pending a change in any of the signals in the process sensitivity list.

Here are some interesting points:

1. If these three assignments were made outside of this process statement, the tools would indicate a "multi-source" or "short" error; however, because these reside within the process and are assigned sequentially, this is not a problem.
2. *Line 3*: *Z* does NOT take on the value of signal *a* at this point! Rather, the value of signal *a* is queued to be placed into *Z* when the process suspends.
3. *Line 4*: as with line 3 *Z* does NOT take on the value of signal *b* at this point! The value of signal *b* now overrides the value of signal *a* in the queue to assign to signal *Z* when the process suspends.

[8] Yes, I know, there is propagation time which impacts the upper limit of the performance of an FPGA and is listed in the static timing analysis report, and there is also the resistance within the wire which causes some of the voltage to be lost as heat, and there are high-frequency harmonics in the transitions that make the traces behave like little transmission lines, but, hey, if you want to get into this level of implementation, go take one of Xilinx's classes!

4. *Line 5*: as with the previous two lines, Z does NOT assume the value of signal c, but now the value of signal c is placed in the queue overriding both previous assignments.
5. *Line 6*: the process does NOT suspend here – Z still has the same value that it did on line 3. Control jumps to line 1.
6. *Line 1*: the process sensitivity list causes the process to enter suspension. At this point all the signals that were assigned to (in this example, only signal Z) takes on the value last assigned to it (in this example, the value of signal c). The process remains suspended until any one or more signals in the process sensitivity list changes value.

Let's examine a simple process to determine the output so that the differences between signals and variables are clearly understood:

```
 1: confuse1:  process (a, b, c)
 2:       variable alpha, beta, gamma  : integer range -50 to 50;
 3:    begin
 4:       alpha := a;
 5:       b     <= alpha;
 6:       beta  := b;
 7:       gamma := c;
 8:       c     <= beta;
 9:       c     <= alpha + beta;
10:    end process confuse1;
```

Snippet 4.15: What Does This Do?

Assume that a, b, and c are defined as integers which may range from −10 to 10. At the time that this process is encountered…

Signal Name	Value
a	−10
b	0
c	10

Spec 4.1

Code analysis:

Line 1: process *confuse1* is defined with a process sensitivity list of *a, b,* and *c* (all assumed to be of type std_logic).

Line 2: variables *alpha, beta,* and *gamma* are defined as range-limited integers and are valid only over −50 <= alpha, beta, gamma <= 50.

Line 3: end of process declarative portion.

Line 4: the variable *alpha* takes on the value of signal *a* immediately.

Line 5: the variable *alpha* is queued for assignment to signal *b*. Signal *b* remains unchanged at this point.

Line 6: variable *beta* takes on the value of signal *b* immediately.

Line 7: variable *gamma* takes on the value of signal *c* immediately.

Line 8: signal *c* is queued with the value of *beta* but does not take on that value until the process enters suspension.

Line 9: signal *c* is queued with the value of *alpha* + *beta* which replaces the previous queued value.

Line 10: control loops back to line 1.

Line 1: signals take on their new assignments.

Let's look at this again, but this time with numbers. Just to clear up a notational issue, any value enclosed by parentheses is considered "queued" for assignment to the signal when the process enters suspension. Table 4.1 represents the status of the signals and variables through the first pass.

Table 4.1 First pass through process *confuse1*.

Line #	*a*	*b*	*c*	*alpha*	*beta*	*gamma*
initial	**−10**	**0**	**10**	**?**	**?**	**?**
4	−10	0	10	**−10**	?	?
5	−10	0 **(−10)**	10	−10	?	?
6	−10	0 (−10)	10	−10	**0**	?
7	−10	0 (−10)	10	−10	0	**10**
8	−10	0 (−10)	10 **(0)**	−10	0	10
9	−10	0 (−10)	10 **(−10)**	−10	0	10
10	−10	0 (−10)	10 (−10)	−10	0	10
1	−10	**−10**	**−10**	−10	0	10

Now here's the interesting thing – notice that the values of signals *b* and *c* have changed! This forces the process to immediately exit suspension and we head through the process again (see Table 4.2).

Table 4.2 Second pass through process *confuse1*.

Line #	*a*	*b*	*c*	*alpha*	*beta*	*gamma*
initial	−10	**−10**	**−10**	−10	0	10
4	−10	−10	−10	−10	0	10
5	−10	−10	−10	−10	0	10
6	−10	−10	−10	−10	**−10**	10
7	−10	−10	−10	−10	−10	**−10**
8	−10	−10	−10 (**−10**)	−10	−10	−10
9	−10	−10	−10 (**−20**)	−10	−10	−10
10	−10	−10	−10 (−20)	−10	−10	−10
1	−10	−10	**−20**	−10	−10	−10

Well, at the time that we reach the process statement again, signals *a* and *b* haven't changed, but signal *c* has! Guess what? The process immediately exits suspension and we go around again! (See Table 4.3.) Finally we've reached a point where the input signals *a, b,* and *c* are no longer changing and the process enters suspension and won't change until any of the signals in the process sensitivity list are changed outside of this process.

Table 4.3 Third pass through process *confuse1*.

Line #	*a*	*b*	*c*	*alpha*	*beta*	*gamma*
initial	−10	−10	**−20**	−10	−10	−10
4	−10	−10	−20	−10	−10	−10
5	−10	−10	−20	−10	−10	−10
6	−10	−10	−20	−10	−10	−10
7	−10	−10	−20	−10	−10	**−20**
8	−10	−10	−20 **(−10)**	−10	−10	−20
9	−10	−10	−20 **(−20)**	−10	−10	−10
10	−10	−10	−20 (−20)	−10	−10	−10
1	−10	−10	−20	−10	−10	−10

Remember that a process is a concurrent statement and that although the statements in a process are "executed" sequentially (that is, in a top-to-bottom order), all of these sequential statements occur simultaneously. Confusing, yes?

Here's how the tools typically handle it:

The synthesis tool attempts to simplify and reduce the sequential statements into combinatorial logic. Variables can be "computed" and "substituted" as the process is analyzed. From the previous process, line 4 simply stores the value of signal *a* into variable *alpha* which is then used on lines 5 and 9. Variable *gamma* is never used at all and would be discarded. Further analysis shows that the first signal assignment to *c* (line 8) is immediately overwritten on line 9 and so, it too, would be discarded (optimized away). If we were to rewrite the process using the substitutions shown, we might see this:

```
entity UART_baudRateGenerator is
  port (reset                : in  STD_LOGIC;
        clock                : in  STD_LOGIC;
        baudRateEnable       : out STD_LOGIC;
        baudRateEnable_x16   : out STD_LOGIC
       );
end entity UART_baudRateGenerator;

architecture BEHAVIORAL of UART_baudRateGenerator is
   begin
   end architecture BEHAVIORAL;
```

Snippet 4.16: Optimized confuse1 Process.

A final level of analysis would reveal that signal *b* will always wind up with the value of signal *a* and therefore this would generate an adder that added signal *a* with itself or a multiply by two. An especially efficient synthesis tool might even recognize that a multiply by two is the same as a wire shift and no logic would be required to implement this process – just a slight renaming of wires as signal *a* is applied to signal *c*.

Simulation tools generally don't run through this type of analysis for simulation, they may actually "loop" through the process. Each one of these loops takes time, but in simulation, these are instantaneous occurrences. The name given to these "zero-time" loops is called a "delta cycle". Referring back to our previous detailed analysis (Tables 4.1–4.3), each table refers to a delta cycle. Most simulation tools will allow for a certain number of delta cycles to pass before it issues a warning that there may be an infinite loop or oscillation occurring.

Consider the following simple example:

```
makeOsc: process (osc)
  begin
     osc <= not osc;
  end process makeOsc;
```

Snippet 4.17: Oscillator (do NOT use in your code!).

Which produces the circuit shown in Figure 4.3.

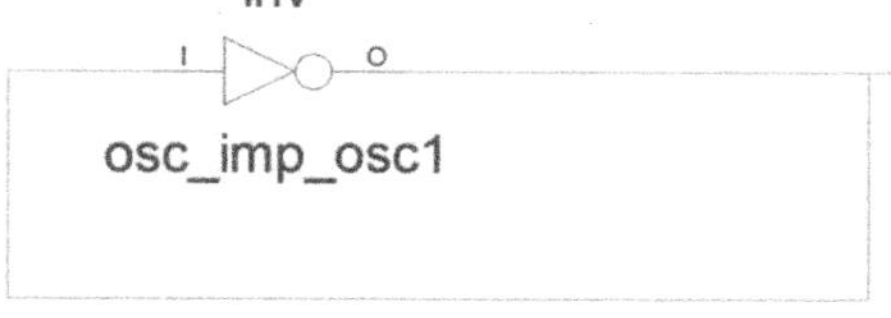

Figure 4.3: Oscillator Schematic (Do NOT Use in Your Code!).

And in simulation, using the ISim simulation tool from Xilinx, the following message is generated:

> "ERROR: at 0 ps: Iteration limit 10000 is reached. Possible zero delay oscillation detected where simulation can not advance in time because signals can not resolve to a stable value in File "C:/test/bookCode/UART_design/latchInference.vhd" Line 60. Please correct this code in order to advance past the current simulation time."

Notice the following aspects of the above message:

1. "Iteration limit of 10,000 is reached" – this is the number of delta cycles that the simulator allows before deciding that there is an infinite loop and stopping. Most simulators, including ISim, have a mechanism for overriding this limit and setting it higher, which is necessary for some processes to resolve.
2. "Possible zero delay oscillation detected where simulation can not advance in time because signals can not resolve to a stable value..." we've been busted! The ISim tool accurately found our oscillator and since the value could not be resolved (despite the

passing of 10 K delta cycles), the simulation could not progress past 0 ps. In short, no simulation time has passed even though many delta cycles have been executed.

Asynchronous vs. Synchronous Processes

I'd be remiss if I didn't mention Asynchronous vs. Synchronous processes. Notice that all of the processes that we've discussed and examined so far have no association with any clock signal. This was done to both introduce the delta-cycle loop and to avoid a premature introduction to the ***if*** statement.

An Asynchronous process, sometimes referred to as a combinatorial process, is simply any process that does not reference a clock. Synchronous processes, by contrast do.

Synchronous processes offer many advantages of asynchronous processes. First, synchronous processes only have a single member in the process sensitivity list – the clock. This avoids many issues regarding having to decide which signals to add to the list and which to omit. If there are any mistakes and a signal is left out of the process sensitivity list, then there is the possibility of a simulation mismatch – where the simulation doesn't behave as the synthesized design does.[9]

Synchronous processes automatically insert registers into the data path. While this may slightly increase latency of a signal through the system it will usually improve system performance (that is, allow for a higher frequency clock) and ease the place-and-route phase of the implementation tools. One might argue that it consumes additional flip-flops, which is true; however, if the flip-flops (located adjacent to the LUTs) are not connected to the output of the LUTs, they go unused, so they are effectively "free".

Best Design Practices 4.2 – Tip

Use synchronous processes whenever possible.
Many FPGA instructors and designers have found that students and clients who view an FPGA as a "sea of flip-flops interspersed with combinatorial logic" rather than "a ton of combinatorial logic contaminated by registers" tend to build faster performing designs and spend less time in debug.

The synchronous process

The Synchronous Process is similar to a combinatorial process with one major difference – the presence of a clock. This affects the process in two places: first, the process sensitivity list

[9] Many tools will indicate a signal if omitted from the process sensitivity list, but it still requires the designer to read the detailed synthesis and simulation reports.

includes one and only one member – the clock. Second, there is a statement within the body of the process that detects either the rising or falling edge of the clock.

Here is the general form for a synchronous process:

```
1: syncProcess: process (clk)
2:  begin
3:     syncEvents: if rising_edge(clk) then
4:        <statements executed on the rising edge of clk>
5:     end if syncEvents;
6:   end process asgn1;
```

Snippet 4.18: The Synchronous Process.

Line 3: if a rising edge is detected on the *clk* signal then all the statements between lines 3 and 5 will be executed.

Line 5: marks the end of the if block.

IEEE provides two functions in the STD_LOGIC_1164 package – rising_edge and falling_edge. These functions detect changes in the clock signal and determine which edge it is. Both functions return a Boolean type which enables to be used in the conditional clause of the if statement (which will be covered in detail shortly).

Resetting processes

Starting from a known good state is always a good idea. Most FPGAs will power-up in a known state with the flip-flops being set to a '0' if not explicitly initialized to a '1'. This is often sufficient for many processes.

Some processes, however, need to be periodically reset when a specified event occurs. These resets can be handled synchronously or asynchronously.

Here are three examples:

```
asyncResetEx: process (reset, a, b, c)
      variable alpha, beta : std_logic := '1';
  begin
     if (reset = '1') then
        alpha := '0';
        beta  := '1';
     else
        <sequential statements>
     end if;
  end process asyncResetEx;
```

Snippet 4.19: Example of a Fully Asynchronous Process with Reset.

```
 1: asyncResetEx2: process (reset, clk)
 2:       variable alpha, beta : std_logic := '1';
 3:    begin
 4:       if (reset = '1') then
 5:          alpha := '0';
 6:          beta  := '1';
 7:       elsif rising_edge(clk) then
 8:          <sequential statements>
 9:       end if;
10:    end process asyncResetEx;
```

Snippet 4.20: Example of a Synchronous Process with an Asynchronous Reset.

```
 1: syncResetEx: process (clk)
 2:       variable alpha, beta : std_logic := '1';
 3:    begin
 4:       syncEvnts: if rising_edge(clk) then
 5:          if (reset = '1') then
 6:             alpha := '0';
 7:             beta  := '1';
 8:          else
 9:             <sequential statements>
10:          end if;
11:       end if syncEvnts;
12:    end process asyncResetEx;
```

Snippet 4.21: Example of a Fully Synchronous Process with Reset.

The first example (Snippet 4.19) shows that when the *reset* signal is asserted, then variable *alpha* is driven low while variable *beta* is driven high. Notice that this differs from the power-on settings of both variables being set high.

When signal *reset* de-asserts, the statements between the "else" and the "end if" statements are executed.

The second example (Snippet 4.20) is very similar except that the "else" statement on line 7 is replaced by and "elsif" (else-if) statement which checks for the rising edge of signal *clk*. If a rising edge is seen then the sequential statements between lines 7 and 9 are executed.

With this example, what happens if the *reset* signal de-asserts scant picoseconds ahead of the clock edge? This can lead to a meta-stable condition in which it is unclear which clock cycle the process de-asserts on.

The third reset example places the *reset* test inside the test for the rising edge of the clock. This means that even if the reset is active, but the clock hasn't yet risen, the process will remain inactive. More to the point, it means that the process will always be aligned with the clock when the *reset* signal is de-asserted, avoiding certain meta-stability problems.

This third example is preferred as it better aligns to best design practices and is more efficient to implement in modern FPGA architecture.

Best Design Practices 4.3 – Tip

1. If it makes sense to NOT have a reset for a process, then do so.
2. If a reset must be used, it should use positive logic ('1' = reset, '0' = operational).
3. If a reset must be used, make it synchronous to the established clock.

4.1.3 *Sequential Statements*

So what are these <sequential statements> that we've been talking about for the last few pages or so? They are the same things that we've been doing with concurrent statements, but with a different syntax.

Let's follow the same basic path that we did when discussing concurrent assignments. First, the unconditional assignment:

```
<target signal> <= <statement>;
<target variable> := <statement>;
```

Snippet 4.22: Syntax for Unconditional Assignment for Signals and Variables.

Just as we've seen from previous examples, whatever is on the right-hand side of the assignment operator must be of the same type and size as what is on the left-hand side of the assignment operator.

Signal assignments are performed using the "<=" operator, and variable assignments are done with the ":=" operator. The details of when the assignments occur are outlined earlier in this section.

Now on to something a little more exciting – conditional assignments. Recall that there are two forms for concurrent signal assignment – the when/else structure and then with/select structure. The sequential domain has the equivalent of these two structures: the if/then/else and the case statements.

If/then/elsif/else/end if

The when/else concurrent structure is used where there may be overlapping conditions or when the assignment may be predicated on a number of different variables or signals. This corresponds to the sequential "if/then/else" statement. Here's the general form:

```
[<label> :] if <condition> then
                <sequence of statements>
          {[elsif <condition> then
                <sequence of statements> ]}
           [else
                <sequence of statements> ]
           end if [<label>] ;
```

Snippet 4.23

Where

<label> is a unique name built using the same rules as an identifier
<condition> is any expression which can be resolved into a Boolean
<sequence of statements> is a list of sequential statements which may include other if statements

If the optional <label> is used with the "end if", then it must match the label name at the beginning of the if clause.

There may be any number of elsif clauses (0 or more). The else clause is optional, but there may only be one.

Let's revisit one of our examples from the concurrent section 3.29:

```
 1: whichQuad: process (angle, radius)
 2:    begin
 3:       if     ((angle >=   0) and (angle <=  90)) then
 4:          target <= UPPER_RIGHT;
 5:       elsif ((angle >=  90) and (angle <= 180)) then
 6:          target <= LOWER_RIGHT;
 7:       elsif ((angle >= 180) and (angle <= 270)) then
 8:          target <= LOWER_LEFT;
 9:       elsif ((angle >= 270) and (angle <= 360)) then
10:          target <= UPPER_LEFT;
11:       elsif (radius < 10) then
12:          target <= BULLSEYE;
13:       end if;
14:    end process whichQuad;
```

Snippet 4.24: Example of the IF/Then Statement.

Code analysis:

Line 1: process *whichQuad* is defined with a process sensitivity list which includes *angle* and *radius*.

Line 3: *angle* is tested to see if it is greater than or equal to 0 and less than or equal to 90. If it is, then target is assigned the value UPPER_RIGHT (on line 4) and control falls through to line 13.

Line 5: *angle* is tested to see if it is greater than or equal to 90 and less than or equal to 180. As with our concurrent case, *angle* can never equal 90 at this point, for if it had, target would have been assigned on line 5.

Line 11: since *angle* is defined between 0 and 360, all of the possibilities are covered and lines 11 and 12 will never be reached.

Line 13: closes the "if" statement.

Notice the absence of the final "else" statement. Since all the possibilities for *angle* are covered, there is no need for one. Also note that this was written as an asynchronous process to better match the example shown in the concurrent section.

Warning! It is generally a bad practice to not include an "else" statement if all possible cases are not covered. Within this final "else" clause should be an assignment to target. Without this clause, if none of the if statements evaluate to logic true, then a latch will be inferred. Due to the asynchronous nature of a latch, they should be avoided as it can have a strong negative effect on the performance. See Best Design Practices 4.3.

Notice the formatting. Certainly VHDL doesn't care about the formatting, but humans do. While the example presented above is one preferred by the author as it provides a columnar view to the conditions and is easy to read, it is certainly not the only "standard" formatting scheme. Many companies suggest or demand that their designers follow specific coding guidelines so that code can easily (comfortably) be read by many designers.

Case statement

Returning to drawing similarities between concurrent statements and sequential statements, the other conditional assignment statement that we saw was the "with/select" statement. This type of comparison structure allows for a single signal to be tested against a series of non-overlapping conditions. The sequential statement that most closely matches this is the case statement.

Here's the general form:

```
[<label>:] case expression is
  when <choices> => <sequential statements>
{[when <choices> => <sequential statements>] }
 [ when others => <sequential statements>; ]
end case [<label>];
```

Snippet 4.25: General Form of the Case Statement.

Continuing with comparing and contrasting the concurrent statements with the sequential statements, we can rewrite Snippet 4.25 within a process as:

```
genSigOut: process (target)
  begin
     pickSigOut: case (target) is
        when UPPER_RIGHT => sigOut <= X"AAA";
        when LOWER_RIGHT => sigOut <= X"555";
        when LOWER_LEFT  => sigOut <= X"861";
        when UPPER_LEFT  => sigOut <= X"79E";
        when BULLSEYE    => sigOut <= X"606";
        when others      => sigOut <= X"000";
     end case pickSigOut;
  end process genSigOut;
```

Snippet 4.26: Example of Case Statement in a Process.

Line 1: process *genSigOut* is defined with a process sensitivity list containing only signal *target*.

Line 3: the case statement is labeled *pickSigOut* and identifies signal *target* as the item to compare against various values specified in the following "when" clauses.

Line 4: if signal *target* is equal to UPPER_RIGHT, then the signal *sigOut* is assigned the value of "1010_1010_1010" or X"AAA" and the case statement is exited, otherwise, the next "when" clause is evaluated.

Lines 5–7: signal *target* continues to be compared against the various values listed in each "when" clause. If the comparison matches, then signal *sigOut* is assigned the value indicated in that line. If a match is found, the case statement is exited.

Line 8: If this process is used in conjunction with the previous if/then/else example, it would be optimized out by the synthesis tools since BULLSEYE is never assigned to signal *target*.

Line 9: if there are additional enumerated types that are valid for *target* then, when each of the listed comparisons fails, this statement would be reached and *sigOut* will be set to all zeros.

Line 10: marks the end of the case statement.

Line 11: marks the end of the process.

Note that both the if/then/else and case statements are shown in their combinatorial (asynchronous) form so that they most closely match their concurrent statement counterpart.

Let's cause a bit of a problem…What would happen if we omitted the "others" clause from the case statement and *target* had several additional enumerated values than what is covered in the when clauses? Certainly if *target* has the value UPPER_RIGHT, LOWER_RIGHT, LOWER_LEFT, UPPER_RIGHT, or BULLSEYE, then the case statement would make the assignment *sigOut* and there are no worries. What happens if *target* is set to MISS?

With the "others" clause present, *sigOut* would be set to all zeros and our discussion would end here. If the "others" clause is NOT present, then how does *sigOut* behave? *SigOut* should retain its last value, but retaining its last value implies that there is something that remembers what the last value was. Since this is a combinatorial (asynchronous) process, the synthesis tools will infer a latch which will accept the new or current value of *sigOut* as long as one of the "when" clauses decodes, but reject any value if none of the "when" clauses decodes.

Why is this an issue? Why the additional discussion? Because this type of coding can ruin FPGA performance!

System performance is calculated based on how fast a signal can move from point 'A' to point 'B'. Since a latch is not associated with a clock and inputs to the gate of the latch may be asynchronous, it is virtually impossible to accurately determine the performance of this type of device. This type of problem is known as "inadvertent latch inference" and can easily be avoided by either using synchronous processes (in which the latches become registers) or covering all possibilities for both the case and if/then/else statements (using a "when others" or a final "else" clause).

Best Design Practices 4.4 – Tip

To achieve the fastest and most reliable performance, avoid using latches! It is important to use the "when others" clause if all possibilities of the case statement are not expressed – especially in an asynchronous process.
Corollary #1: Use registers instead
Corollary #2: completely cover all possible cases in both "if/then/else" and "case" statements when using asynchronous processes!
Make every effort to avoid using asynchronous processes!

Let's complete our UART design by attacking the lower-level modules and implementing these modules with what we've just covered – processes.

We'll begin by first constructing the baud rate generator module. We left off having defined the entity for this module:

```
entity UART_baudRateGenerator is
   port (reset                : in  STD_LOGIC;
         clock                : in  STD_LOGIC;
         baudRateEnable       : out STD_LOGIC;
         baudRateEnable_x16   : out STD_LOGIC
        );
end entity UART_baudRateGenerator;
```

Snippet 4.27

Now we can build the architecture associated with this entity:

```
entity UART_baudRateGenerator is
  port (reset                : in  STD_LOGIC;
        clock                : in  STD_LOGIC;
        baudRateEnable       : out STD_LOGIC;
        baudRateEnable_x16   : out STD_LOGIC
       );
end entity UART_baudRateGenerator;

architecture BEHAVIORAL of UART_baudRateGenerator is
   begin
   end architecture BEHAVIORAL;
```

Snippet 4.28

Code analysis:

Lines 1–7: the entity statement which was described in previous chapter.

Line 9: the architecture is defined and named "BEHAVIORAL" and is associated with the above-listed entity. Note that the name "BEHAVIORAL" is arbitrary; however, it is often named after the coding style used (such as "BEH", "STRUCTURAL", "RTL", etc.). Other useful names for the architecture might refer to the FPGA family that it is targeting, or a specific algorithm used in the architecture, so that just in case there are multiple architectures associated with this entity, it would be relatively easy to know what each architecture was intended for.

Line 10: ends the architecture's definition region and marks the beginning of the circuit description.

Line 11: ends the architecture.

We know that this module should produce two enable signals – one occurring at the desired baud rate and one at 16 times the baud rate. For now, let's assume that we want to use 115,200 baud and that the system clock is running at 50 MHz. Doing the math, we realize that 50M/115,200 = 434.02777. Fortunately, the RS-232 specification does not require precise timing and we can simply produce one enable pulse for every 434 clocks. Continuing to show our prowess in math, we can compute the oversample clock rate at 50M/(115,200 * 16) = 27.126736, so one enable pulse will be produced for every 27 clock pulses. Since RS-232 is an asynchronous protocol, the two enable pulses are not required to be aligned in time. This simplifies our design and we can now write two processes that are synchronous to the system clock, but otherwise not necessarily related to each other.

We'll start by creating the two processes within the architecture:

```
 1: . . .
 2: architecture BEHAVIORAL of UART_baudRateGenerator is
 3:    begin
 4:       make_x16en: process (clock)
 5:          begin
 6:             syncEvents: if rising_edge(clock) then
 7:
 8:             end if syncEvents;
 9:          end process make_x16en;
10:
11:       make_baudEn: process (clock)
12:          begin
13:             syncEvents: if rising_edge(clock) then
14:
15:             end if syncEvents;
16:          end process make_baudEn;
17:
18:    end architecture BEHAVIORAL;
```

Snippet 4.29

The contents of each of these processes will be very similar – they will both contain a counter, a comparator, and a single-pulse generator…

```
 1: . . .
 2: architecture BEHAVIORAL of UART_baudRateGenerator is
 3:    begin
 4:       make_x16en: process (clock)
 5:             variable clockCount : integer range 0 to 31 := 0;
 6:          begin
 7:             syncEvents: if rising_edge(clock) then
 8:                baudRateEnable_x16 <= '0';
 9:                clockCount := clockCount + 1;
10:                isCountDone: if (clockCount = 17) then
11:                   baudRateEnable_x16 <= '1';
12:                   clockCount := 0;
13:                 end if isCountDone;
14:             end if syncEvents;
15:          end process make_x16en;
16:
17:       make_baudEn: process (clock)
18:             variable clockCount : integer range 0 to 511 := 0;
19:          begin
20:             syncEvents: if rising_edge(clock) then
21:                baudRateEnable <= '0';
22:                clockCount := clockCount + 1;
23:                isCountDone: if (clockCount = 434) then
24:                   baudRateEnable <= '1';
25:                   clockCount := 0;
26:                end if isCountDone;
27:             end if syncEvents;
28:          end process make_baudEn;
29:
30:    end architecture BEHAVIORAL;
```

Snippet 4.30

Continuing the discussion from Snippet 3.30.

Process make_x16en:

Line 8: baudEnable_x16 is driven inactive. Unless this variable is set to another value later in the process, this value will "stick" when the process suspends (in synthesis, this will occur on the next rising clock edge).

Line 9: variable *clockCount* is incremented. Since *clockCount* is a variable, it will take on the incremented value immediately so that it can be tested on the following line.

Line 10: variable *clockCount* is tested against 27 – the value that we calculated to produce a sample rate of 16 ∗ 115,200

Line 11: if *clockCount* equals 27, then signal *baudEnable_x16* is driven high on process suspension (change in clock state).

Line 12: *clockCount* is reset to 0.

Line 13: end of test for count completion.

Line 14: end of "true" statements to be executed on the rising edge of the clock.

Lines 21–27: this code is identical to the *make_x16en* process except that *baudEnable* is the target signal and that the enable is generated once per 434 clocks instead of once per 27 clocks.

This is a good place to have a brief discussion about aspect of coding style. Because of VHDL's ability to assign a value to a signal in more than one place within a process and that that signal takes on the last signal assignment prior to the process suspending, this enables us to "list out" the signals and their "default" values prior to performing specific tests, rather than having to list each signal and its default value in each test block.

For example, if we have a case statement with many signals such as…

```
bunchaWhens: process (clk)
  begin
    syncEvents: if rising_edge(clk) then
       case (someIntegerSelector) is
          when 0 =>
             sigA <= '0';
             sigB <= '0';
             sigC <= '0';
             sigD <= 'Z';
             sigE <= '1';
          when 1 =>
             sigA <= '0';
             sigB <= '0';
             sigC <= '0';
             sigD <= '1';
             sigE <= '0';
          when 2 | 3 =>
             sigA <= '0';
             sigB <= '0';
             sigC <= '1';
             sigD <= 'Z';
             sigE <= '0';
          when 4 to 10 =>
             sigA <= '0';
             sigB <= '1';
             sigC <= '0';
             sigD <= 'Z';
             sigE <= '1';
          when others =>
             sigA <= '1';
             sigB <= '0';
             sigC <= '0';
             sigD <= 'Z';
             sigE <= '0';
        end case;
     end if syncEvents;
  end process bunchaWhens;
```

Snippet 4.31

Notice that in the example above only one of the signals (*sigA–E*) changes in any given state. Listing each signal explicitly is beneficial from the standpoint that each "when" statement can be investigated on its own, but at the expense of longer processes and the lack of clarity as to which signal is being changed (which can easily be overcome with a comment).

Many designers choose to put the "default" values for signals ahead of the case or if statements and list only the signals that change.

```
 1: shorterBunchaWhens: process (clk)
 2:   begin
 3:     syncEvents: if rising_edge(clk) then
 4:        sigA <= '0';
 5:        sigB <= '0';
 6:        sigC <= '0';
 7:        sigD <= 'Z';
 8:        sigE <= '0';
 9:        case (someIntegerSelector) is
10:           when 0 =>
11:              sigE <= '1';
12:           when 1 =>
13:              sigD <= '1';
14:           when 2 | 3 =>
15:              sigC <= '1';
16:           when 4 to 10 =>
17:              sigB <= '1';
18:              sigE <= '1';
19:           when others =>
20:              sigA <= '1';
21:         end case;
22:      end if syncEvents;
23:   end process shorterBunchaWhens;
```

Snippet 4.32: Listing "Default" Values Ahead of Specific Drives.

Code analysis:

Lines 4–8: signals *sigA* through *sigE* are given initial values. As always, since these are signals, they don't take on these values immediately – only when the process enters suspension.

Line 9: the case statement is entered and based on the integer value of *someIntegerSelector* one (and only one) of the following when statements will be reached.

Line 10: if *someIntegerSelector* has the value of zero, then signal *sigE* is set to receive the value of '1' when the process suspends. Since this is the only line of code associated with this when clause, control flows to the "end case".

Line 12: if *someIntegerSelector* has the value of one, then control passes to line 13, otherwise control passes to the next when clause on line 14.

Line 13: signal *sigD* is queued to receive the value of '1'.

Line 14: case statements support the use of lists which allow for more than one explicit value to be checked. In this when statement if *someIntegerSelector* has the value of either 2 or 3, then control transfers to line 15.

Line 16: as was just mentioned, the case statement supports the use of ___ and here we see the range test. If *someIntegerSelector* has the value of 4, 5, 6, 7, 8, 9, or 10, then control passes to the next statement, which is on line 17.

Lines 17 and 18: when clauses may (and usually do) include more than one statement. In this clause *sigB* is queued to receive the value of '1' and *sigE* is queued to receive the value of '1'.

Line 19: if none of the explicitly listed when clauses are matched, the "when others" clause is accepted. All statements between this point and the end of the case statement are "executed".

Line 21: end of the case statement.

Remember that this technique is not limited to case statements, but can be used with if/then/endif and if/then/elsif/end if statements…

```
 1: defaultIfEx: process (clk)
 2:   begin
 3:      syncEvents: if rising_edge(clk) then
 4:         <list of default conditions>
 5:         ifBlk: if (<condition1>) then
 6:            . . .
 7:         elsif (<condition2>) then
 8:            . . .
 9:         else
10:            . . .
11:         end if ifBlk;
12:      end if syncEvents;
13:   end process defaultIfEx;
```

Snippet 4.33: Outline of How Listing "Default" Values Ahead of if Statement Can Be Used.

Code analysis:

Line 4: this line represents multiple lines in which the "default" conditions are listed.

Line 5: represents the beginning of the series of tests for various conditions. If <condition1> evaluates to a Boolean "true" then the statements between this point and the next "elsif" or "else" or "end if" are "executed", otherwise control is transferred to the next test which is on line 7.

Line 7: if <condition2> evaluates to a Boolean "true" then the statements between this point and the next control statement are executed.

Line 8: if neither <condition1> nor <condition2> are "true" then the statements between the else and the "end if" are "executed".

You may be wondering why these "default" values can't simply be listed out in the "else" clause of the "if" statement or the "when others" clause in the "case" statement. The answer is that if this is the only location where the values are set to their default conditions, then the user must guarantee that this statement is reached prior to any of the other cases being reached and that it must be the first state reached, otherwise the values will be undefined until enough states have been reached or until the else clause is reached.

Coding the transmitter

Let's continue the design with the transmitter as it is a bit simpler than the receiver to understand. We know that the transmitter must send a start bit followed by 8 bits of data, followed by a stop bit. Each bit must be sent at a specific time which is dictated by the signal *baudRateEnable*.

From this information (and a bit of experience) we can see that a simple Moore-type state machine can be employed to solve this problem quickly and accurately.

SIDEBAR 4.2

State Machines

A full discussion of state machines is outside the scope of this book, but we can rehash the important points.

There are basically two types of state machines - a "Mealy"-type state machine which combines the current state with combinatorial logic to produce the outputs and determines what the next state will be. "Mealy"-type state machines tend to have fewer states than the "Moore" state machines, but because of the use of combinatorial logic to form the outputs, this type of state machine is more difficult to properly simulate (in that there are more conditions that must be validated), and tend to have a lower performance in FPGAs.

The "Moore"-type state machine tends to have more states, with each state producing one, and only one, set of outputs. This makes the "Moore"-type state machine very easy to validate in simulation and performs better in the register-rich FPGA environment.

The state machine will need to reside in the IDLE state which must keep the serial output line driven high (see Figure 4.4). When the request is made to begin the transmission, then the state will advance to SEND_START_BIT which will drive the serial output line low for one baud rate period of time. Once this bit is sent, the 8 data bits will be transmitted in the SEND_DATA_BITS state. When all 8 bits are sent, then the state machine advances to the SEND_STOP_BIT state where the serial output line is driven high for one baud rate time. Finally, once the stop bit is sent, the device returns to the IDLE state where it waits for the next transmit request.

What might be the best way to implement a state machine? Consider that only one variable or signal is being examined – that of the current state. Many designers choose the "case" type statement to implement Moore-type state machines.

While there are a number of different approaches to construct state machines, the conceptually easiest one is that of the "single-process" state machine. It is thus named in that only one process is used to manage both the determination of the next state and the generation of the output for the current state.

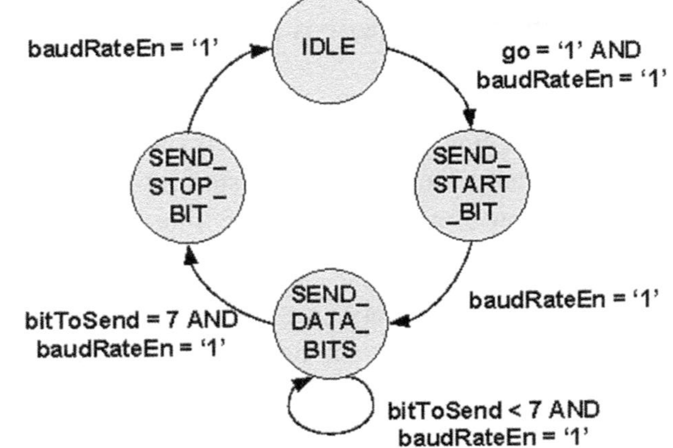

Figure 4.4: State Machine for the RS-232 Transmitter State Machine.

Other state machine designs may use "two-process" state machines which use one process to determine the next state (usually implemented as a synchronous state machine), and the other to drive the current outputs for the state (often implemented as an asynchronous state machine).

Since this UART design is not going to tax the capabilities of the FPGA (that is, there is more than enough performance and resources inherent in the FPGA to achieve the desired baud rates without having to use more complex coding techniques), we will only consider the single-process state machine.

Let's first look at the overall design of the transmitter state machine. Since all activity is tied to the main clock in the FPGA, this can (and should) be coded as a synchronous process:

```
tx_sm: process (clock)
   begin
     syncEvents: if rising_edge(clock) then
        . . .
      end if syncEvents;
   end process tx_sm;
```

Snippet 4.34: Outline for the Transmitter State Machine Process.

Should this process be resettable? Ideally, coding a process so that a reset is not required can reduce the effort that the tools need to make in order for a design to meet the user-specified performance requirement. Without a reset, this process would merely finish transmitting any character it had already started to send, then just wait for a new character. If a reset were used, then the current character would be aborted and the line would return to logic high (as dictated by the RS-232 specification) and whatever was receiving the transmitted data would likely receive a bad character.

Best Design Practices 4.5 – Coding Resets

Whenever possible, write code so that it is self-resetting so that no external reset is required.
Corollary – apply resets to modules whose outputs may produce undesirable effects downstream or when self-resetting will prevent the process from reaching a usable state before an upstream source begins providing data.
Best Coding Practices: always code for active high, synchronous resets as these best fit current FPGA architectures (i.e. no additional resources are required and the highest performance will be achieved).
Corollary: some FPGA silicon resources require active low resets. Code the active low reset as close to the silicon resource instantiation/inference as possible.

Since the baud rate for this design is fairly high (115,200 baud), there is very little "human" wait time, so the need to abort a character midway is low. If this design were slated for a much lower baud rate, then it might make sense to abort the currently transmitting character.

So, for a "real-world" design, this process could be coded without a reset and be expected to perform quite adequately. However, since this is a little "academic" we'll code in the reset anyway, just so that you can see how it is done…

```
 1: tx_sm: process (clock)
 2:    begin
 3:      syncEvents: if rising_edge(clock) then
 4:         resetRun: if (reset = '1') then
 5:            -- signals and variables to be reset
 6:         else
 7:            -- operational description
 8:         end if resetRun;
 9:       end if syncEvents;
10:    end process tx_sm;
```

Snippet 4.35: Outline for the Transmitter State Machine Process with Reset Coded.

Notice that the labels for both lines 3 and 4 are used on lines 8 and 9 to match the "end if" with the appropriate "if". While labels are not required, they can be very convenient for readability (consider multiply nested if and case statements where the "end if" or "end case" might be dozens of lines previous and careful indentation might not have been performed – it might make understanding or debugging the code considerably more difficult. Additionally, any error or warning messages related to an "if" or "case" statement will be referred to by its label thus making debugging easier).

Let's start coding what we already know we're going to need. We know that we are going to have a state machine that contains four states. First we must create a user-defined enumerated type to describe the states. Next we will need a signal or variable to contain the actual state.

So, should the object that contains the actual representation of the state be a signal or variable? If it were a signal, then the state would only change when the process suspended. Since this is a synchronous process (i.e., tied to the clock which is in the process sensitivity list), this is an acceptable choice. Additional benefits of coding the state signal is that many diagnostic tools such as simulators and in-system debuggers such as Xilinx's ChipScope Pro will have access to the signal making this easier to debug. As a signal, the current state may be viewed by other processes or other concurrent states in this module.

Certainly a variable can be used within the process and since variables take on their values immediately, then if the "next" state needed to be tested within the process, then this would be the only way to go. Defining this as a variable also limits the "scope" to the current process, so if you had multiple processes within this module, then each process could have its own variable named "txState" which would be independent of each other.

Since this is a simple module, defining the state object as a variable or signal has effectively no impact, so for this example, we'll define txState as a signal so that we have visibility to it using the simulation tools and any "in-chip" debuggers.

From a light examination of the state machine, all the transitions from state to state are based on the baud rate enable signal. It would be a simple task to place an "if" statement around the state machine so that each state doesn't have to test for the baud rate enable signal, effectively "factoring out" the enable.

Next, we also know that there is going to be a signal, possibly only one clock period wide, marking when we should begin transmitting the data. We also know that our state machine is only going to be active when the baud rate enable signal is asserted, which may not align when the signal to begin transmission is issued. It would be a good idea to capture both the data to transmit and the start transmission signal and hold it until a baud rate enable is seen. The implication here is that the transmission of the data might be delayed by a baud rate period.

At this point we can code the following:

```
 1: tx_sm: process (clock)
 2:    begin
 3:       syncEvents: if rising_edge(clock) then
 4:          resetRun: if (reset = '1') then
 5:             -- signals and variables to be reset
 6:              txState       <= IDLE;
 7:              serialDataOut <= '1';
 8:              go            := '0';
 9:
10:          else
11:             -- capture the transmitRequest and data
12:             catchStart: if (transmitRequest = '1') then
13:                go       := '1';
14:                dataToTx := parallelDataIn;
15:             end if catchStart;
16:
17:             -- TX state machine (conditioned by baudRateEnable)
18:             smEnabled: if (baudRateEnable = '1') then
19:                sm: case (txState) is
20:                   when IDLE =>
21:                   when SEND_START_BIT =>
22:                   when SEND_DATA_BITS =>
23:                   when SEND_STOP_BIT =>
24:                end case sm;
25:             end if smEnabled;
26:          end if resetRun;
27:       end if syncEvents;
28:    end process tx_sm;
```

Snippet 4.36: Outline for the Transmitter State Machine Process Partially Coded.

Code analysis:

Lines 5–8: reset block filled in. Notice that the signals are given initial values to drive the outputs into a quiescent and inactive state.

Lines 11–15: the transmitRequest signal (which may be very long or only a one clock wide pulse) is captured in case it doesn't align with the baudRateEnable assertion. Since parallelDataIn may only be valid during the transmitRequest, it too is captured for later use in the state machine.

Lines 18 and 25: everything between these two lines is only active when baudRateEnable is asserted.

Lines 19–25: the outline of the case statement for the state machine. Notice the absence of the "others" statement. This is appropriate since each of the four possible states is tested for. If an "others" statement is included, many synthesis tool will report a warning indicating that the "others" statement will not be reached and will be optimized out.

Let's code the four states of the state machine one at a time…

The IDLE state must hold the transmit line at a high level as defined by the RS-232 specification. Since the state machine is not transmitting, but rather just waiting for the command to start, it must also assert the *ready* signal so that any external module that is monitoring this signal will know that the transmitter is ready to begin transmitting.

The IDLE state must also check the *go* variable to see if the *transmitRequest* has or had been asserted since the last *baudRateEnable* was active. If it was, then the state machine must advance to the next state which will drive the line low. While it is possible, and even one baud time more efficient, to drive the line low in this baud rate time, this state is split out for clarity.

Since the process to begin transmitting has begun, the *ready* signal is de-asserted so that other modules will know not to make a *transmitRequest*.

The *go* variable is cleared at this point (although it can be reset anywhere else in the state machine prior to re-entering the IDLE state), as is the counter which tracks the number of bits sent.

The IDLE state now looks like this:

```
1: when IDLE =>
2:    ready         <= '1';
3:    serialDataOut <= '1';
4:    timeToStart: if (go = '1') then
5:       go         := '0';
6:       bitToSend:= 0;
7:       ready     <= '0';
8:       txState   <= SEND_START_BIT;
9:    end if timeToStart;
```

Snippet 4.37: The Transmit State Machine IDLE State.

Code analysis:

Line 1: tests *txState* against **IDLE**.

Line 2: signal *ready* is driven high to indicate to the other modules that the transmitter state machine is ready to receive new data to transmit.

Line 3: signal *serialDataOut* is driven high as this is required by the RS-232 standard as an inactive condition (i.e., when no data are being transmitted, this is how the line should be set).

Line 4: if the *go* variable was set either in this clock cycle or any clock cycle since the last *baudRateEnable* then lines 5 through 8 will be evaluated.

Line 5: the *go* variable is cleared in anticipation of the next *transmitRequest* assertion. This could just as easily be performed in any other state, the most likely being the last state prior to re-entering the IDLE state.

Line 6: variable *bitToSend* is reset to 0. This variable is used to identify which of the bit positions will be placed on the output signaling line.

Line 7: since the process of transmitting the *parallelDataIn* has begun, this process is no longer able to receive additional data. A more sophisticated UART might use a First-In

First-Out (FIFO) to buffer multiple requests issued over a short period of time. If this were the case, the FIFO's "full" flag would be inverted and used as the *ready* signal (that is, the FIFO/serial transmitter is "ready" as long as there is room in the FIFO for more data). The *transmitRequest* signal would then be used as a write enable into the FIFO (which would then require the *transmitRequest* signal to be only one clock wide otherwise characters would be written into the FIFO as long as the *transmitRequest* was active). Finally, the *go* variable would be replaced by the FIFO's inverted empty signal (that is, if there is still data in the FIFO, then the transmitter would then send the next character, otherwise it would wait in the IDLE state until more data are placed into the FIFO).

Line 8: the state machine, upon receiving the next *baudRateEnable* assertion, will undergo transition to the next state which will send the start bit. Certainly the SEND_START_BIT could be incorporated into this state by merely driving the output line low and transitioning to the SEND_DATA_BITS state; however, it is coded as shown to keep more closely to the way that the state machine diagram is drawn.

Line 9: ends the statements to evaluate if variable *go* were active.

The SEND_START_BIT state is much simpler as it only requires the output line to be driven low for one baud rate period, then transition to the state in which the data bits are transmitted.

```
1: when SEND_START_BIT =>
2:    serialDataOut <= '0';
3:    txState       <= SEND_DATA_BITS;
```

Snippet 4.38: The Transmit State Machine SEND_START_BIT State

SEND_DATA_BITS state must perform two basic tasks: first, determine which bit of the captured data should be sent, and second, decide to select another bit to transmit or to exit to the final state.

```
1: when SEND_DATA_BITS =>
2:   serialDataOut <= dataToTx(bitToSend);
3:   whenDone: if (bitToSend = 7) then
4:      txState <= SEND_STOP_BIT;
5:   else
6:      bitToSend  := bitToSend + 1;
7:   end if whenDone;
```

Snippet 4.39: The Transmit State Machine SEND_START_BIT State.

Line 2: we'll define the variable *dataToTx* as a std_logic_vector of 8 bits (7 downto 0). The variable *bitToSend* then becomes the index into the std_logic_vector and effectively "points" to which bit is to be sent next. Whichever bit is being pointed to is then driven onto *serialDataOut*.

Line 3: since we are sending 8-bit data which starts at index 0, then when *bitToSend* is 7, there is no more data that need to be sent and line 4 is evaluated. Otherwise any other value of *bitToSend* that is less than 7 will cause the statements associated with the "else" clause to be evaluated (line 6).

Line 4: when all the bits have been sent, then the state is advanced to the SEND_STOP_BIT state which will be reached on the next assertion of *baudRateEnable*

Line 6: if there are more bits to be sent, then the variable *bitToSend* is incremented. Upon the next occurrence of *baudRateEnable* Line 2 will be reached as there is no change in state and the next bit value will be placed on the signaling line.

Generally, a state machine which must be coded for performance should NOT include additional logic such as the adder, but rather enable a counter located in a separate process. This process will be revisited later in the book and will implement both the FIFO and a higher-performing state machine (although at 115,200 baud, it is certainly not required).

The final state, SEND_STOP_BIT must issue a line idle signal (high level) for at least one baud period. Certainly the RS-232 protocol supports other stop bit durations, but this is a simple design and these other values won't be supported.

In addition to the line idle, the SEND_STOP_BIT will remain in its current state until the *transmitRequest* is de-asserted. If this is not done and the module which requested the transmission does not release the *transmitRequest* signal, then control would return to the IDLE state and immediately begin retransmitting whatever data were present on the *parallelDataIn* signal.

```
1: when SEND_STOP_BIT =>
2: serialDataOut <= '1';
3: if (transmitRequest = '0') then
4:    txState <= IDLE;
5: end if;
```

Snippet 4.40: The Transmit State Machine SEND_STOP_BIT State.

Code analysis:

Lines 3 and 5: notice that no label has been used. This is legal and is often done for very small "if" statements as there is little likelihood of the code being misunderstood.

The only thing left now is to define any signals that we've used when coding this process. This leaves us with the completed module looking like this...

```
library IEEE;
use IEEE.STD_LOGIC_1164.ALL;
entity UART_transmitter is
    port ( reset           : in  STD_LOGIC;
           clock           : in  STD_LOGIC;
           baudRateEnable  : in  STD_LOGIC;
           parallelDataIn  : in  STD_LOGIC_VECTOR (7 downto 0);
           transmitRequest : in  STD_LOGIC;
           ready           : out STD_LOGIC;
           serialDataOut   : out STD_LOGIC
                 );
end entity UART_transmitter;

architecture Behavioral of UART_transmitter is
      type legalStates is (IDLE, SEND_START_BIT,
                             SEND_DATA_BITS, SEND_STOP_BIT);
      signal txState : legalStates := IDLE;
   begin
      tx_sm: process (clock)
         variable dataToTX : std_logic_vector(7 downto 0);
         variable bitToSend: integer range 0 to 7 := 0;
         variable go       : std_logic := '0';
      begin
         syncEvents: if rising_edge(clock) then
            resetRun: if (reset = '1') then
            txState       <= IDLE;
            ready         <= '0';
            go            := '0';
            serialDataOut <= '1';
         else
            catchStart: if (transmitRequest = '1') then
               go := '1';
               dataToTx := parallelDataIn;
            end if catchStart;
            smEnabled: if (baudRateEnable = '1') then
               sm: case (txState) is
                  when IDLE =>
                     ready         <= '1';
                     serialDataOut <= '1';
                     timeToStart: if (go = '1') then
                        go        := '0';
                        bitToSend:= 0;
                        ready     <= '0';
                        txState   <= SEND_START_BIT;
                     end if timeToStart;
                  when SEND_START_BIT =>
                     serialDataOut <= '0';
                     txState <= SEND_DATA_BITS;
                  when SEND_DATA_BITS =>
                     serialDataOut <= dataToTx(bitToSend);
                     whenDone: if (bitToSend = 7) then
                        txState <= SEND_STOP_BIT;
                     else
                        bitToSend  := bitToSend + 1;
                     end if whenDone;
                  when SEND_STOP_BIT =>
                     serialDataOut <= '1';
                     if (transmitRequest = '0') then
                        txState <= IDLE;
                     end if;
               end case sm;
            end if smEnabled;
         end if resetRun;
       end if syncEvents;
    end process tx_sm;
end architecture Behavioral;
```

Snippet 4.41

Now let's tackle the receiver portion of the design. This is a little trickier as RS-232 is defined as an asynchronous protocol and we don't know when the character will arrive relative to the baud rate enable or even with respect to the much faster clock.

For the moment we'll assume that the arriving data is synchronized with the clock so that we don't need to contend with meta-stable issues and can focus on the design.

Our oversampling clock will take 16 samples at the desired baud rate (Figure 4.5 shows only 8 samples per bit due to graphic limitations). Since *serialDataIn* is asynchronous, it may or may not align with *baudRateEn_x16*. The basic strategy of this design is to detect the falling edge of *serialDataIn* which marks the beginning of the start bit. Since we know that the best place to sample the incoming data is in the middle of the bit and that there are 16 samples per bit, we will wait 8 samples (a half bit) before re-sampling the start bit to insure that a line glitch doesn't accidentally begin the receive character process. Once the start bit is accepted, every 16th sample will be aligned to the middle of the *serialDataIn* bit. Each 16th bit can then be shifted into receive buffer. Once the full number of bits are received, the STOP bit can be sampled to ensure proper packet formatting and the data can then be issued along with a *dataValid* pulse to indicate that the received data are valid.

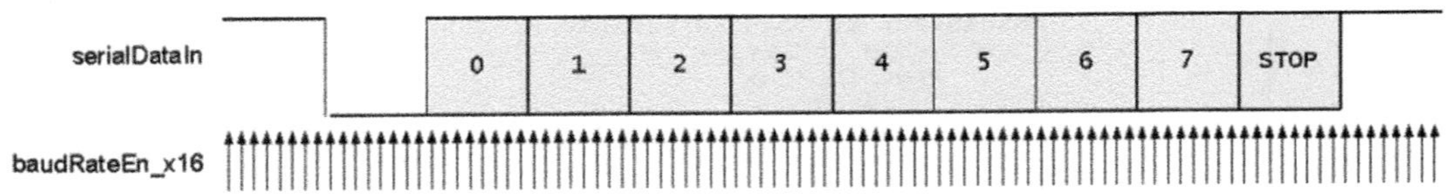

Figure 4.5: RS-232 Protocol and the Oversampling Pulse (8 of the 16 per bit are shown).

There are several ways of handling *dataValid*. Some designers prefer to issue a single clock period pulse while others prefer to hold the *dataValid* signal active until a new character is seen on the *serialDataIn* signal. This design will use the former strategy so that any component downstream of this UART will only need to detect the presence of the *dataValid* signal and not have to add an extra check to see when *dataValid* de-asserts.

Notice that there is a lot more counting going on than with the TX state machine (see Figure 4.6). We can still count the number of bits to receive the same way we counted the number of bits to transmit in the TX state machine, but we'll count the oversample enables using another process and control it with a few basic signals. This approach is more applicable to designs which must operate at higher speed and is demonstrated here so that you can apply this technique to your own higher-speed designs.

As we did with the transmit state machine, we'll outline the process first, then backfill each major aspect of the module.

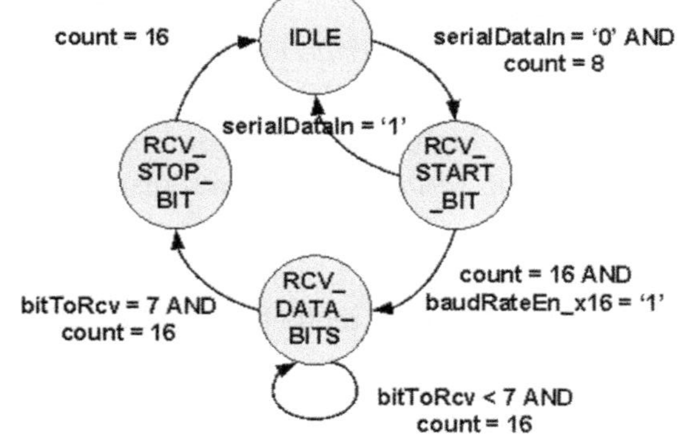

Figure 4.6: State Machine Bubble Diagram for the RS-232 Receiver.

```
rx_sm: process (clock)
   begin
      syncEvents: if rising_edge(clock) then
         resetRun: if (reset = '1') then

         else
            smEnabled: if (baudRateEnable_x16 = '1') then
               sm: case (rxState) is
                  when IDLE =>
                  when RCV_START_BIT =>
                  when RCV_DATA_BITS =>
                  when RCV_STOP_BIT  =>
               end case sm;
            end if;
         end if resetRun;
      end if syncEvents;
end process rx_sm;
```

Snippet 4.42: UART Receiver State Machine Process

Starting with the IDLE state, we know that it doesn't have to do anything except to keep the outputs inactive, load the counter to count for 8 *baudRateEnable_x16*s, and transition to the RCV_START_BIT when the arriving data are seen to transition low.

```
when IDLE =>
   countEnable <= '0';
   startDetect: if (serialDataIn = '0') then
      bitsReceived:=  0;
      countValue  <=  7;
      countEnable <= '1';
      countLoad   <= '1';
      rxState     <= RCV_START_BIT;
   end if startDetect;
```

Snippet 4.43: IDLE State for the UART Receiver Process.

Code analysis:

Line 2: holds *countEnable* inactive until *serialDataIn* is at zero. A marginally better design is to create a temporary variable that holds the last value of *serialDataIn* so that the state machine "activates" when the actual high-to-low transition is seen rather than assuming that the line is high and that a low level on *serialDataIn* represents a falling edge. This assumption disappears at the first pass through the state machine as the final state in the state machine insures that the line is "live" before returning to look for the start bit.

Line 3: tests if serialDataIn has dropped, indicating the beginning of the start bit.

Lines 4–8: initializes the values required to run the state machine. Notice that *countValue* is set to 7 in order to find the middle of the start bit.

RCV_START_BIT state's job is to re-sample the start bit and if that re-sampling shows that the line has returned to a high level (indicating that the transition from the IDLE state was a glitch), then return control of the state machine back to the IDLE state. If the line has stayed at a low level, then it is a valid start bit, the counter should now be set to count 16 oversample enables and transfer control to the REC_DATA_BITS state.

```
 1: when RCV_START_BIT =>
 2:   isHalfWay: if (countDone = '1') then
 3:      stillStartBit: if (serialDataIn = '0') then
 4:         countValue <= 15;
 5:         countLoad  <= '1';
 6:         rxState    <= RCV_DATA_BITS;
 7:      else
 8:         rxState    <= IDLE;
 9:      end if stillStartBit;
10:   end if isHalfWay;
```

Snippet 4.44: RCV_START_BIT State for the UART Receiver Process.

Code analysis:

Line 2: *countDone* is driven by the counter process and when the *countValue* (currently set to 7) is complete, signal *serialDataIn* is re-sampled. If it is low, then lines 4 through 6 are evaluated, otherwise the code assumes that there was a glitch and returns control to the IDLE state and awaits another start bit.

The RCV_DATA_BITS state is very similar to the SEND_DATA_BITS state from the transmitter state machine. As each count of 16 expires (placing the sample in the middle of the serial bit) the sample is stored in one of eight positions in parallelDataOut according to the value in *bitsReceived*. Once all 8 bits have been received, control advances to the RCV_STOP_BIT state.

```
when RCV_DATA_BITS =>
   countEnable <= '1';
   collectingOrDone: if (countDone = '1') then
      parallelDataOut(bitsReceived) <= serialDataIn;
      if (bitsReceived = 7) then
         rxState <= RCV_STOP_BIT;
      else
         bitsReceived := bitsReceived + 1;
      end if;
      countLoad <= '1';
   end if collectingOrDone;
```

Snippet 4.45: RCV_DATA_BITS State for the UART Receiver Process.

The final process, RCV_STOP_BIT, samples *serialDataIn* when the count from 16 has expired. The stop bit is defined to be level high. If this high level is seen, then it is assumed that the serial data are valid and the *dataValid* signal is asserted. If the high level on the serial line is not seen, then there is some type of problem which could be not enough bits of data transmitted, baud rates that don't match, or even a cable disconnect. In this event, dataValid is never asserted and control remains in this state until the line returns to a high level at which time control is passed to IDLE.

```
when RCV_STOP_BIT  =>
   sampleStopBit: if (countDone = '1') then
      countEnable <= '0';
      isStopBit: if (serialDataIn = '1') then
         dataValid <= '1';
         rxState   <= IDLE;
      else
         lineDown <= '1';
      end if isStopBit;
   end if sampleStopBit;

   isLineDown: if (lineDown = '1') then
      checkLineLevel: if (serialDataIn = '1') then
         lineDown <= '0';
         rxState  <= IDLE;
      end if checkLineLevel;
   end if isLineDown;
```

Snippet 4.46: RCV_STOP_BIT State in Receiver Process.

This completes our basic UART design. Our next step is to verify the behavior of each of the modules, then of the entire design.

Here's a quick review comparing processes and process statements with architecture and some concurrent statements.

Table 4.4 Summary

	Architecture	Process
Information in/out	An architecture uses an entity to enumerate the number and type of information. Architectures only have access to the signals defined in the entity statement and in the architecture - they can't "see" variables within processes	Processes have access to all the signals available within the architecture - both those provided in the entity statement and those defined within the architecture. Additionally, any constants, variables, types, subtypes, attributes, etc. that are defined within the process are limited to within the process in which they are defined
Conditional assignment	The concurrent form for a general conditional assignment is the "when" structure	The sequential form is more akin to the software structure: if/then/else structure
Conditional assignment	The concurrent form for a selected conditional assignment is the "with/sel" statement	The sequential form is more akin to the software structure "switch/case"

Coding the Test Bench for the Top-Level UART Design

At the end of the last iteration, we left off with a partially complete test bench for testing the entire UART design. Following Best Design Practices, each module should be simulated independently before; however, all the important aspects of the simulation can be shown in this one example, so in the interest of time and paper these have been omitted from the book.[10]

We are left with the design for the test bench (see Figure 4.7).

Now that we've covered processes, let's code the data generator.

First, we need to list out what the data generator does (the behavioral specification). It must take a list of predetermined characters and send them one by one to the *parallelDataIn* port of the UART transmitter. Once applied, it must assert the *transmitRequest* signal for at least one clock cycle. It must then wait for the transmitter to assert the *txIsReady* signal before repeating this cycle until there are no more characters to transmit.

The previous processes that we've examined needed to be synthesizable so we limited ourselves to constructs that can be physically realized. Since this process only needs to operate in the simulation domain, we have a bit more freedom with the way we code.

[10] Rest assured that these modules were independently simulated and verified before the final test bench was written!

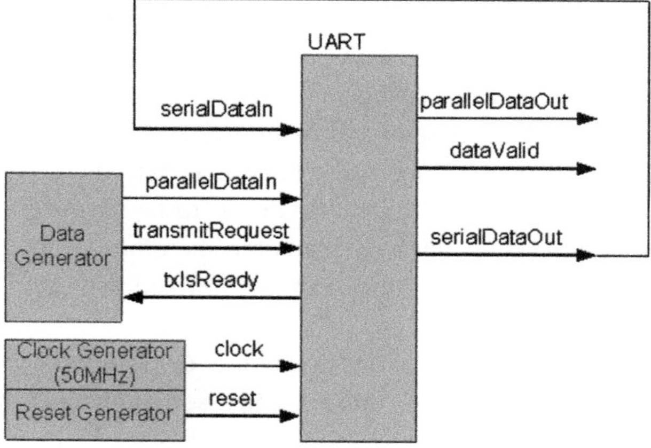

Figure 4.7: Simple Test Bench for the UART Design.

Here's what we know so far:

1. A message of one or more characters must be transmitted.
2. Characters from this message must be converted from a character to a std_logic_vector, which is what the *parallelDataIn* requires.
3. The *transmitRequest* can only be asserted when *txIsReady* is high and can be anywhere from at least 1 clock period wide to infinity. From the code, we know that *txIsReady* will only go high when *transmitRequest* has returned low.

Let's code this process up…

```
mkParallelIn: process
     variable message : string (1 to 15) := "This is a test!";
     variable cCharToSend : character;
     variable iCharToSend : integer range 0 to 255;
   begin
     wait until (reset = '0');
     wait for clock_period * 10;
     sendMsg: for i in 1 to 15 loop
        cCharToSend     := message(i);
        iCharToSend     := character'pos(cCharToSend);
        parallelDataIn  <= std_logic_vector(
                              to_unsigned(iCharToSend,8));
        wait until (txIsReady = '1');
        transmitRequest <= '1';
        wait for clock_period * 5;
        transmitRequest <= '0';
      end loop sendMsg;
      wait;
  end process mkParallelIn;
```

Snippet 4.47: The Data Generator for the UART design.

Code analysis:

Line 2: variable *message* is defined as a 15-character-long string. Notice how the low index begins at 1 – this is a requirement for strings.

Lines 3 and 4: variables *cCharToSend* and *iCharToSend* are defined as a single character and an integer between 0 and 255, respectively. Note that both these variables represent the same piece of information, namely, the character that is to be passed to the transmitter, however, their forms are different – one is a character, the other an integer. These share the same name (as they represent the same thing), but are prefixed with a 'c' and an 'i' to indicate the type of information they contain.

Line 6: the process suspends until the reset signal is de-asserted. This is coded in this fashion so that should the amount of time that reset is asserted change, this will automatically adjust so that the stimulus that this module produces will never be asserted during the initial reset period.

Line 7: this line provides 10 times the clock period additional delay after the de-assertion of the reset signal. Technically this is not needed, however, there are designs that have modules that don't come out of reset immediately. Consider a memory controller that, on reset, clears the entire memory. This process may consume hundreds or thousands (or more) clock cycles to completely clear the memory. If this were the case with this design, the * 10 would be increased to slightly more than the maximum number of clock cycles required by the memory process to clear the memory.

Line 8: since there are 15 characters in the *message*, this loop will iterate 15 times – once per character.

Line 9: the specific character to be transmitted is extracted from *message* and placed in *cCharToSend.*

Line 10: the character extracted in the previous step is converted into an integer. The 'pos is a predefined attribute which we will cover in the final loop of this book.

Line 11: the integer equivalent of the character is converted into a std_logic_vector and applied to the input of the Unit Under Test (UUT).

Line 12: the process re-enters suspension waiting for *txIsReady* to assert. If *txIsReady* is already asserted, then this line takes 0 time.

Line 13: Now that we know that the transmitter is ready, the *transmitRequest* is brought high.

Line 14: the process re-enters suspension for five times the clock period which is more than enough time for the transmitter to detect the *transmitRequest* assertion. Certainly this value can be anything from 1 clock period to a very large period of time.

Line 15: after the designated amount of time specified on line 14 has passed, the *transmitRequest* is de-asserted.

Line 16: control transfers back to line 8 while *i* is less than 15, otherwise control transfers to the following line.

Line 17: As the data have been exhausted, the process enters permanent suspension and never reactivates. The only way to reactivate this process is to restart the simulation.

The partial waveform view of Xilinx's ISim Simulation tool is shown in Figure 4.8. It is possible to squint and inch our way through the waveforms to check the results of the simulation. This can become fairly tedious and error prone. We'll revisit this design in our last iteration to construct a test bench which produces text messages which are easier and clearer to see and understand (see Figure 4.8).

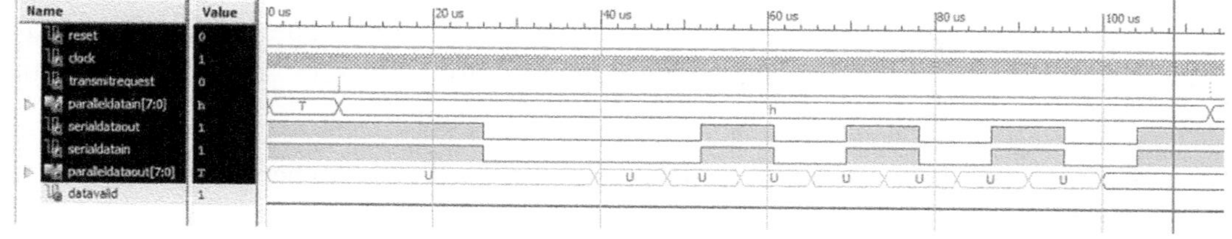

Figure 4.8: Simulation of UART Design.

4.2 Tool Perspectives – Synthesis Options and Constraints

While this section isn't strictly about VHDL, it is important to have a basic understanding in order to implement a successful FPGA design. Early in the book we made a distinction between how an FPGA design *behaves* vs. its *performance*.

Synthesis options do not change the behavior of the code, but it can have significant effect on how that behavior is achieved within the FPGA. The synthesis tools are responsible for making choices as to *how* the VHDL design is converted into a netlist.

Certainly when FPGA design primitives are instantiated, the synthesis tools have little or no effect, but for other coding styles where there is any inference at all, the synthesis tools can and will have significant impact.

4.2.1 Synthesis Options

The synthesis options specify styles, preferences, and direction to the synthesis engine dictating how the netlist is to be constructed. Recall that most VHDL analyzers are capable of inferring how code is interpreted. The synthesis options help the tool decide how to encode state machines, construct memory, make speed vs. area tradeoffs, replicate registers to reduce fan-out, and other decisions.

Please read your vendor's Synthesis User Guide to see what options are offered and suggestions for their best use.

Although the default synthesis options for most tools are adequate for simplistic designs, they are generally not sufficient for designs which have higher performance requirements. Following good coding practices and intelligent synthesis options can produce quality netlists.

There are a number of ways of controlling or selecting the synthesis options. The easiest method is to set the synthesis options globally at the top level of the design. This will apply the options uniformly to all modules within the design. The downside is that while generally satisfactory, there are a number of situations where it can produce less than optimal code.

One commonly used option is that of fan-out. Simply put, fan-out is the number of lines that must be driven by a single source. Physics being what it is, tells us that the more wires that are driven, the higher the capacitance, the longer it takes to drive. So, one "trick" is to limit the number of lines that are driven by duplicating the source signal.

Note that more resources are used to lower the fan-out to achieve higher performance (see Figure 4.9).

This does have a downside. If the fan-out is limited to 10 signals, but 100 lines are required, the original source signal will be replicated 10 times. This is one of many mechanisms the tools use to improve performance at the expense of silicon real-estate.

Now consider the situation where a design has a small portion of a much larger design allocated to high-speed processing (requiring higher performance) and a much larger section for lower clock rates. If the synthesis options (tailored to meeting the performance of the high-speed portion) were applied uniformly to the entire design, there could be a completely unnecessary "explosion" of replicated logic, possibly causing the design to not fit into the selected FPGA and always consuming more power than is required.

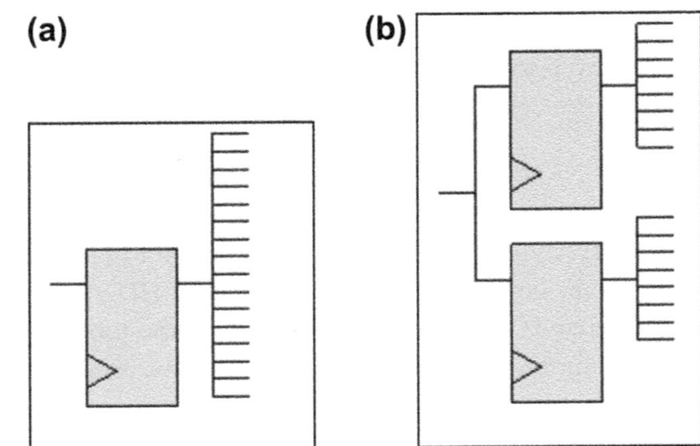

Figure 4.9: (a) High Fan-Out Condition (Left) and (b) Lower Fan-Out Condition (Right).

This class of problem can be avoided by specifying specific synthesis options via a special constraint file or as attributes within the VHDL. Either of these mechanisms will override the top-level options set in the Graphical User Interface (GUI) for the synthesis tool for a specific module thus enabling the user to tailor the synthesis options to the requirements for each design element.

Some designers prefer to take a more "manual" control over their design environment by running the synthesis tools in the command line mode and applying a specific set of synthesis options for each module and synthesizing each module separately. Remember that VHDL is a "bottom-up" language, that is, the lowest levels of the design must be processed (analyzed) before the next layer up in the hierarchy. This is handled automatically by many of today's design environments; however, when "manual" control is taken, the user becomes responsible for proper synthesis order. Many advanced designers employ custom scripts or Makefiles to manage their design or embed synthesis options directly into their source code.

Table 4.5

Option	What It Does
Fan-out	Sets the replication level for signals so that no signal source is required to drive more than the specified number of signals. This keeps the drive capacitance low
Finite state machine encoding	Tools may select One-Hot, compact, Gray, or other encoding scheme to balance performance vs. resources
Keep hierarchy	Keeping hierarchy preserves hierarchical information which is beneficial when analyzing or debugging the design. Keeping the hierarchy removes some of the tools ability to optimize the code if best design practices haven't been followed
Memory resources	The tools will run a tradeoff analysis between performance and number and type of resources. This is usually between block RAM (relatively limited, but large and fast) against distributed RAM (numerous, but slower)

4.2.2 Constraints

Synthesis constraints are directives to the synthesis tools used to try to meet performance objectives. Recall that the job of the synthesis tools is to convert VHDL files into a netlist which will later be used in the implementation process. Synthesis can approximate delays through basic logic elements, but has no idea about where these logic elements are located or how the path that will ultimately connect them together is routed.

Based on the family and part number that you specify to the synthesis tool, a fan-out delay model, and the synthesis options the synthesis engine will attempt to make decisions that will meet the synthesis constraints.

Note that constraints differ from synthesis options in that the constraints specify WHAT the performance objectives are and the synthesis options specify HOW those performance objectives might be met.

CHAPTER 5

Loop 3

The final loop in this book will refine and extend your understanding of VHDL by introducing the concept and mechanisms of reuse. One key aspect to reuse is the awareness of the existing Institute of Electrical and Electronics Engineers (IEEE) libraries and the commonly used packages within them.

Finally, we will revise the UART test bench to both incorporate reuse concepts and to enhance your ability to write a solid VHDL test bench.

5.1 Introducing Concept of Reuse

Reuse is a much-touted term that has not seen the acceptance that it should. The concept of reuse is simple – build it in a generic fashion once, then use it many times, in many different modules and projects.

Two challenges that have hindered reuse are as follows: first, since many designers are not familiar with the concept from the beginning of their training, their code tends to be very specific to the project that they are working on and not suitable for reuse. This section should help the reader overcome that hurdle! While it is true that writing code in a generic fashion may be slightly slower (especially when one is not practiced at it), however when the code is used again, even only once, this small cost in time is more than compensated for – remember that it is not just the time coding that is saved, but the time verifying and debugging (which in many cases takes considerably longer than the original coding effort!). The second challenge is that today's tools aren't configured to efficiently support code snippets and package/library modules. Unfortunately, until a decent snippet/package manager is written, this responsibility must be left to the user. Suggestions for management are provided toward the end of this section.

Reuse in VHDL is realized in two main fashions: components and packages. Components can be in the form of source code or black boxes. Black boxes are encrypted netlists that can be added into a design, and the source code is not accessible, therefore the black box cannot be modified. Typically, companies that sell their Intellectual Property (IP) will provide their products as an encrypted netlist to minimize the chance of IP theft and to protect their trade secrets from reverse engineering.

Certain tools, such as Xilinx's CoreGen tool, allow the designer the ability to tailor some of the key attributes of a core to meet the specific need of the design. Once these values have

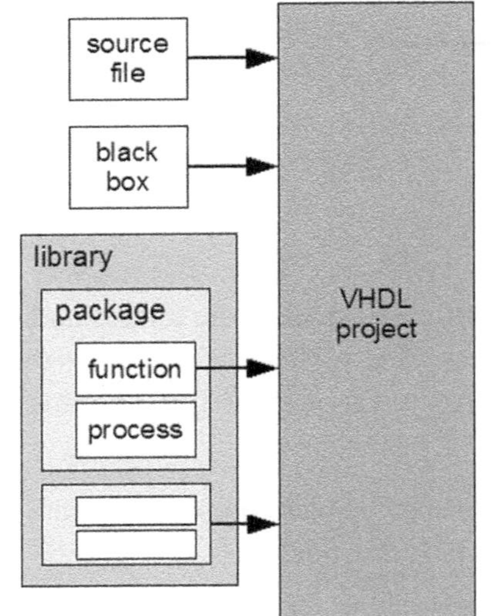

Figure 5.1: Relationship of Reusable Modules.

been selected, a black box is produced and no further changes can be made to that black box without regenerating it (see Figure 5.1).

The process of creating an encrypted netlist for the purposes of delivering secure IP[1] is outside the scope of this book and has more to do with the vendor's synthesis tools than with VHDL.

Libraries, as we have seen before, are comprised of a collection of packages. These packages, in turn, may contain functions, processes, component declarations, constants, and types/subtypes. Behavior in a package is contained within the functions and processes.

Components written or delivered as regular VHDL source code can be coded with *Generic* and *Generate* statements and use Predefined *Attributes*. These statements provide the designer certain flexibility when coding, which can enable the component to have various parameter values set during synthesis and simulation.

5.1.1 A Little More on Libraries and Packages...

We've already seen some examples of libraries and packages usage introduced in the previous loops. Let's take a moment to review...

A package is a specific VHDL construct. A package may define new types and subtypes, constants, component declarations, function declarations, and procedure declarations (which

[1] IP means Intellectual Property. This is a thought or idea that is captured in code. Some examples are RS-232 UARTs, PCIe Root Complex, microprocessors, and DSP filters.

SIDEBAR 5.1

Reuse in "Real-Life"

Reuse is a beautiful thing, but how is it implemented in the "real-world"?

There are a number of companies which produce and sell IP. Their job is to write specific forms of IP for sale to designers that require complex and very specific forms of IP. Typically these "cores" sell for between $5,000US and up with many of the cores residing in the $20,000US -$50,000US range. Some examples of these cores include PCIe DMA controllers and root complexes, specialized video interfaces such as Display Port and DVI. Despite the seemingly large price tag on these cores, it is often less expensive to purchase such a core than it is to develop it "in-house" as it will take several months (if not years) to develop and require a group of engineers who have the detailed knowledge of the core that is to be developed. Consider: $50,000US is only 50 man-days of labor or less than 2 man-months. Factor in verification on top of coding and delay to the project delivery and one can quickly see that the price tag is "cheap" compared to internal development.

Another form of reuse is through the IEEE and other organizations. Typically, the IEEE doesn't provide IP per se, but rather provides collections of packages that support specific tasks. This book relies heavily on the std_logic_1164 and numeric_std packages from the IEEE library in order to manage multi-valued logic and the plethora of conversion and mathematical operations that can be performed on them.

Vendors typically offer their own libraries and packages to support their silicon products. Libraries such as Xilinx's UNISIM and SIMPRIM are collections of silicon "primitives" for the various Xilinx FPGA families and are provided within Xilinx's development environment.

Larger (or more formalized) companies, especially where ASIC and FPGA have been designed for many years, often implement their own internal libraries and generally place these libraries on servers where their employees can access them. These libraries and packages will often have their own test benches associated with them so that they can be verified against new versions of various tools. Formal company processes (such as including documentation and test benches) generally exist prior to a package being added to the collection to ensure consistent use and proper behavior of the packages throughout the company.

Small teams within the company may share "packages" in a more informal fashion, whether it be on an internal network or a "sneaker-net". More often than not, these don't enjoy the formalization process which includes documentation and test benches.

Then there is you. Even if a package is not shared with others, it should still be documented and have its own test bench as you will likely be using it for years! (If you used it once - you'll use it again!)

behave as function prototypes from C/C++). Typically a package is associated with a package body. The "package body" is a secondary construct and cannot exist without a package.

The package body contains only the functions and procedures required to support those functions and procedure definitions declared in the package. The package body may also include functions and procedures, such as "function 'D'" in Figure 5.2, used to support the functions and procedures declared in the package, but not available to the user.

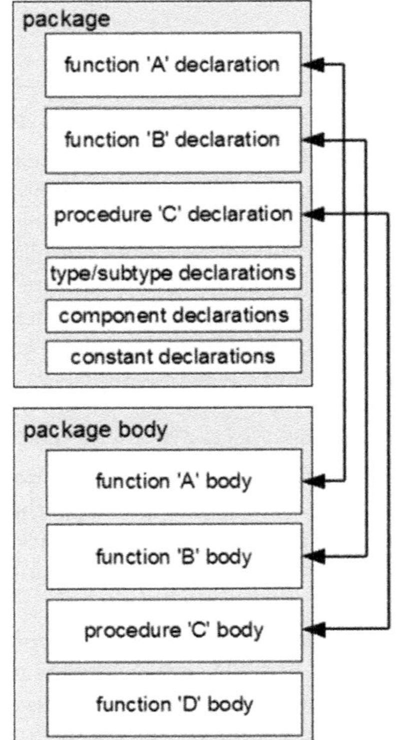

Figure 5.2: Relationship Between a Package and a Package Body.

A library is a collection of packages. The specific means of implementing libraries is not covered in the VHDL specification and left to the vendors to implement as they wish. As a result, the physical structure of the working directory and the libraries may be inconsistent among the various vendors. Some vendors structure libraries as directories and physically place the associated packages into that directory, while other vendors choose to keep all the packages in the same physical directory and group the packages logically into libraries. The user may impose his or her own file system structure on the libraries and convince the tools to access these libraries appropriately.

There are two steps that are required for using a library. The first step is to declare the library for use. The syntax is quite simple:

```
library <library name>;
```

Snippet 5.1

where <library name> is a legal identifier corresponding to an available library. If the library is not available in the workspace the tools will indicate an error and analysis will not continue.

Once the library is announced, the packages within that library become available with the following statement:

```
use <library name>.<package name>.<content specifier>;
```

Snippet 5.2

where:

<library name> is a legal library name that has already been made accessible with the "library" statement,

<package name> is a legal name of a package contained within the library called out in <library name>,

<content specifier> is either the keyword "all" which makes the entire contents of the package available to the design or it may be a single member within the package. A single member may be an individual constant, type, function or procedure definition, etc.

If there are multiple items within a package that are to be used, but not all items in the package, then multiple "use" statements may be utilized to enumerate each desired item. This is often done to avoid conflicts between package members or just desiring to be specific.

More commonly, the entire package is made available by using the keyword "all" as the <content specifier>. When the "all" is specified, every component definition, type and subtype definition, function and procedure declaration, and constant definition can be seen by the module in which it was specified (and ONLY the module in which it was specified). Every VHDL module where the contents of a library/package must be seen must contain the library and all the necessary use definitions.

Allowing access or visibility into a package does not cause the design to run slower or consume more resources. Items in the packages are incorporated into the design only when they are specifically used.

There are two exceptions to the library declaration and use statements. These two libraries that are always available to the designer and do not need to be explicitly declared. They are the STD library which automatically uses the STANDARD package. This package defines integers, reals, bit, bit_vectors, and other basic data types. This package does NOT include the commonly used std_logic/std_logic_vector types.

The other library is named "WORK". This is the default library in which all codes exist when not explicitly placed into another library.

Explicitly listing the WORK and STD libraries is unnecessary, but is legal and should not pose any problems.

Resolving conflicts between packages

Sometimes packages carry duplicate definitions. Perhaps there are constants that are named the same, but have different definitions, or there might be functions and procedures that may have identical names and arguments.

The following example demonstrates a conflict between two packages containing several constants (see Snippets 5.3 and 5.4).[2]

```
package package1 is
  constant K0 : integer := -1;
  constant K1 : integer := 0;
  constant K2 : integer := 1;
end package package1;
```

```
package package2 is
  constant K0  : integer :=  0;
  constant K10 : integer := 10;
  constant K20 : integer := 20;
end package package2;
```

Snippet 5.3: Packages with Some Overlapping Constants.

```
 1: library WORK;
 2: use WORK.package1.all;
 3: use WORK.package2.all;
 4:
 5: entity overlapMain is
 6:   port ( A : out  integer range -10 to 20;
 7:          B : out  integer range -10 to 20;
 8:          C : out  integer range -10 to 20);
 9: end entity overlapMain;
10:
11:architecture Behavioral of overlapMain is
12:   begin
13:      A <= K0;
14:      B <= K1;
15:      C <= K10;
16:   end architecture Behavioral;
```

Snippet 5.4: Both Packages Loaded with "All" Cause Conflict.

Code analysis:

Line 1: the Library WORK is defined. Recall that WORK is one of the two libraries that is automatically declared. This line of code can be omitted with no ill effects.

Line 2: all of the members of Package1 are made available.

Line 3: all of the members of Package2 are made available.

Lines 5–9: describes the interface to the outside world. In this case all the signals are outputs. The integer range will be rendered as a number of wires, in this case, most likely 6 bits – 1 bit for sign, 5 bits to represent the maximum (and most negative maximum) values.

[2] The astute reader may simply ask – "'ell, if they're *constants* shouldn't they always be the same?" If we're talking about physical constants, yes, but what about constants that are used for tiltering or encryption/decryption. In these cases, the constant is only constant for one particular equation or algorithm and may vary from use to use.

Line 13: Which K0 is used? The one from Package1 or the one from Package2? Since VHDL refuses to make any kind of assumption as to what the designer coded, the "compiler" will issue an error indicating that there is a conflict which must be resolved on this line.

One way to resolve this conflict is to call out all the members of the package that don't conflict, or more efficiently, only the members of that package which are used (see Snippet 5.5).

```
library WORK;
use WORK.package1.K1;
use WORK.package2.all;
```

Snippet 5.5: Solution #1 - Limiting Members from Package-1.

Once could also select the individual members from Package2 in the same fashion. This is enough to resolve the issue.

Consider: what if there were a thousand constants in each package and most were used? This would make for an extremely tedious task!

Fortunately, we can create aliases for these items. Aliases are used to rename items for both ease of use and conflict resolution. By explicitly calling out only the non-duplicate names, this conflict can be easily resolved.

We will use the following structure for the alias:

```
alias <unique_name> is <library name>.<package name>.<element name>;
```

Snippet 5.6

In this way, the full name of the item is spelled out and this resolves the naming conflict while providing a meaningful name for ease of use and clarity (see Snippet 5.7).

```
library WORK;
use WORK.package1.all;
use WORK.package2.all;

entity overlapMain is
   port ( A : out  integer range -10 to 20;
          B : out  integer range -10 to 20;
          C : out  integer range -10 to 20);
 end entity overlapMain;

architecture Behavioral of overlapMain is
     alias the_K0_I_really_want is WORK.package2.K0;
   begin
     A <= the_K0_I_really_want;
     B <= K1;
     C <= K10;
  end architecture Behavioral;
```

Snippet 5.7: Using Aliases to Resolve Package Conflicts.

Code analysis:

Line 1: library WORK is declared. This is unnecessary as WORK is always available.
Line 2: all members of package1 are made available for use.
Line 3: all members of package2 are made available for use.
Line 12: a unique name is assigned to the full path from the library, through the package, down to the specific element. Since both packages refer to K0, the tools have no idea which K0 to use and the tools will throw an error. After this alias is defined, the K0 from package2 can be accessed by using the name *the_K0_I_really_want*.

Code snippets

Code snippets are incomplete pieces of logic which represent a frequently used construct. Theoretically, these could be turned into components, functions, or procedures; however, the designer either doesn't want to go through the trouble, or the snippet must be tweaked for the application and there is no easy way of making the code "generic".

Since there is no formality to snippets, it is up to the designer to provide some type of snippet management. The easiest way is to set aside a directory and place the snippet and explanatory code into a text file and name it something recognizable. The snippet can then be cut-and-pasted into a VHDL source file.

5.2 Flexibility Using Generics and Constants

Two similar constructs exist within VHDL that can be used in a similar way – the generic and the constant. While the values of both Generics and Constants can't be changed post-synthesis, they are different constructs intended to serve two distinct purposes.

A constant, as we recall from Loop 1, is a fixed value of a specified type. This constant, as the name suggests, is immutable and fixed throughout its scope. A constant simply takes a value of a specified type and can be assigned to other signals or variables of the same type. It can be of any legal VHDL data type – std_logic_vector, integer, real, or time (just don't expect to synthesize the latter two!), or even a user-defined type. The value or type of this constant can only be changed by editing the source code, resynthesizing, and re-implementing the design.

Constants can be declared within an entity and are valid for that entity and any architecture associating with that entity and therefore valid or "scoped" anywhere within that architecture, or within functions, procedures, and processes (where the constants are only valid within those constructs).

Constant definitions look a lot like signal declarations, except for the keyword being "constant" instead of "signal" and the signal's optional initial value assignment part which is required for the constant (see Snippet 5.8).

```
1: constant period        : time := 10 ns;
2: constant n_columns     : integer := 5;
3: constant initial_state: legal_states := IDLE;
4: constant ground        : std_logic_vector(31 downto 0) := (others=>'0');
```

Snippet 5.8: Examples of Constant Definitions.

Explanation:

Line 1: this is a commonly used construct within a test bench. Notice the use of the "ns" units, which is one of the possible time units required for the physical type of time. Using this coding style, all references to anything having to do with this clock may use the period constant instead of trying to manually compute time. Coding in this way supports test-bench constructs such as "wait for 50∗period;" to cause the test bench to "pause" at this point for 50 times the duration of a clock period. This is particularly useful when the final clock period may not be known or what-if scenarios where different clock periods are being tested.

Line 2: this line of code demonstrates a synthesizable definition for n_columns which is set to an integer value of 5. Notice that the "range" construct is not applicable here since there is no range, only a single value.

Line 3: constants may also be made for user-defined types, in this case, an enumeration type called "legal_states" of which one of these states is named "IDLE". Of course, "legal_states" must already have been defined as a type or subtype.

Line 4: from time to time, you may need to set a signal or a subrange of values to zero. Having a vector of grounds can make this very easy as this definition shows. Notice the use of the "others" statement. This should key you in (as well as the example from line 3) that you can use array aggregates to define a constant, and, extending this concept, you may even have records and arrays which are constant!

Generics on the other hand…

A generic value is constant only within a specific scope or region of code or instantiation. Generics have a specific value and type – just like that of a constant. If a component is "passed" a generic through its entity, then the value of that generic behaves exactly like a constant, but only within that specific instance of that component.

For example, in the code snippet below, if one component that uses generics is instantiated twice and passed different generics, then the generic "width" will be constant within that instantiation (8 in instantiation mod1, 16 in mod2).

(a)

```
. . .
mod1: register_n
    generic map (width => 8)
    port map (a=>byte1, b=>byte_out_1);

mod2: register_n
    generic map (width => 16)
    port map (a=>word2, b=>word_out_2);
. . .
```

(b)

```
entity register_n is
    generic (width : integer := 4);
    port (a : in  std_logic_vector(width-1 downto 0);
          b : out std_logic_vector(width-1 downto 0));
end entity register_n;
```

Snippet 5.9: Generics Example.

Generics can only be declared within the entity of a component and are accessible only within that entity and any architecture associated with that entity.

Values for generics can be "passed" into a component giving that component different qualities or capabilities. For example, as shown in the previous example, a parallel register can be created and the inputs and outputs can be determined by a constant providing the capability of quickly and easily adjusting the size of the parallel register based on the value of the generic.

Since generics can differ from instantiation to instantiation, the component in the example can be a 1-bit register, an 8-bit register, or a 32-bit register, provided the architecture associated with the entity uses the generics and is coded properly.

We'll take a look at the architecture for the *register_n* entity in the generate statement section (next coming up).

Entities can support more than one generic and generics can be of any legal type...

```
entity memory_block is
    generic (DATA_WIDTH      : integer := 4;
             ADDRESS_WIDTH   : integer := 8;
             PARITY_CHECK_ON: Boolean := false
            );
    port (dataIn   : in  std_logic_vector(DATA_WIDTH-1 downto 0);
          address : in  std_logic_vector(ADDRESS_WIDTH-1 downto 0);
          writeEn : in  std_logic;
          dataOut : out std_logic_vector(DATA_WIDTH-1 downto 0)
         );
end entity memory_block;
```

Snippet 5.10: Example of Multiple Generics in the Entity.

Most "compilers" provide the option of specifying the values for top-level generics from the command line (or Graphical User Interface, GUI). In this way, the code doesn't have to be modified for accessing different capabilities within the code.

Generics are also used in a similar fashion as #defines are used in C. The value of the generic can be tested and "conditional" code can be included or excluded when used in conjunction with the generate statement which will selectively produce logic.

5.3 Generate Statements

Generate statements are mechanisms for controlling if or how many times a certain component gets instantiated. There are two forms for the generate statement – "conditional" and "loop/repetitive".

Generate statements are often, as they are here, tied in with the use of generics for the simple reason that if a generic is used as the control to a generate statement (either form), then modules can be quickly adjusted to fit a new application. For example, if a counter module were developed, then a generic statement could be used to specify the operational range of the counter.[3]

5.3.1 Conditional Generate Form

The conditional generate statement is roughly analogous to the #ifdef block structure of C/C++. Software sees that if a certain variable is defined, then the code is compiled (added to the object). In the case of hardware, if the conditional generate statement evaluates to "true" then the code within the conditional generate statement is produced. Keep in mind that this occurs during synthesis (elaboration) and NOT at run-time which means that there is no speed/performance penalty for generating code, only that more silicon resources within the Field-Programmable Gate Array (FPGA) are consumed for synthesis. When generates are used within simulation, no logic is produced, but there is a performance penalty as the computer must simulate the logic that is generated.

In its general form:

```
<label>: if (<condition>) generate
  { <concurrent statements>; }
end generate [<label>];
```

Snippet 5.11: General Form of the Conditional Generate.

where

<label> represents a legal and unique label identifier
<condition> is any expression that results in a Boolean
<concurrent statements> zero or more legal concurrent statements

[3] The advantage to keeping the counter as small as possible is that this saves on logic resources.

Any concurrent statement is legal within the body of the generate condition, the most common, however, are instantiations as shown in Snippet 5.12.

```
 1: library IEEE;
 2: use IEEE.STD_LOGIC_1164.ALL;
 3:
 4: library UNISIM;
 5: use UNISIM.VComponents.all;
 6:
 7: entity generateMain is
 8:     generic (useDifferential : boolean := false);
10:     port ( sigA_p : in  STD_LOGIC;
11:            sigA_n : in  STD_LOGIC;
12:            sigZ_p : out STD_LOGIC;
13:            sigZ_n : out STD_LOGIC
14:          );
15:   end entity generateMain;
16:
17: architecture Behavioral of generateMain is
18:     signal A : std_logic := 'U';
19:     signal Z : std_logic := 'U';
20:   begin
21:     -- build the single-ended buffers
22:     snglBufs: if (not useDifferential) generate
23:       bufA: IBUF port map (I=>sigA_p, O=>A);
24:       bufZ: OBUF port map (I=>Z,      O=>sigZ_n);
25:     end generate snglBufs;
26:
27:     -- build the differential buffers
28:     diffBufs: if (useDifferential) generate
29:       bufA: IBUFDS port map (I=>sigA_p, Ib=> sigA_n, O=>A);
30:       bufZ: OBUFDS port map (I=>Z, O =>sigZ_p,  Ob=>sigZ_n);
31:     end generate diffBufs;
32:
33:     Z <= not A;
34:
35:   end architecture Behavioral;
```

Snippet 5.12: Code Snippet Illustrating Conditional Generates.

Code review:

Lines 4,5: The UNISIM library is a Xilinx proprietary library which contains the component definitions for all the device primitives. In this example, it is included to support the use of input and output buffer instantiation.

Line 8: This generic, in concert with the conditional generates, guides the synthesis tool as to what to construct. This generic can be set from the "command line" or synthesis tool options so that the source code does not need to be modified in order to make a different selection. See your vendor's User Manual to see what the command line option is.

Lines 22–25: this group of statements is used to generate single-ended buffers if the generic *useDifferential* is false (see Figure 5.3). Notice that only the _p signals are used.

Lines 28–31: this group of statements is used to generate differential buffers if the generic *useDifferential* is true (see Figure 5.4). VHDL version 2008 supports more advanced conditional generate statements which would allow this to be written as an if/else rather than two if statements with opposite tests.

Line 34: simple logic to tie the input and output buffers together. This is normally where the user logic would reside.

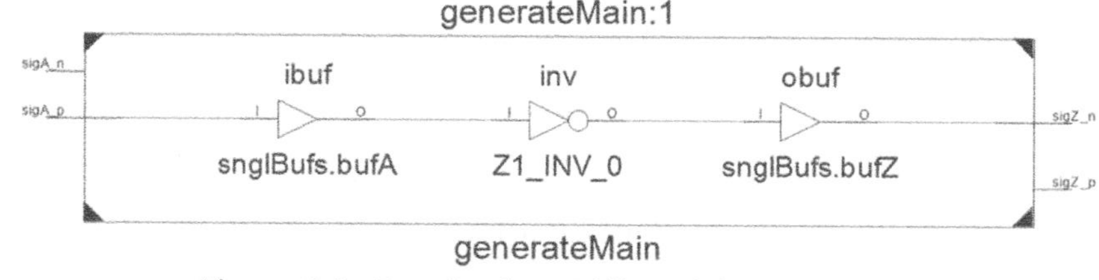

Figure 5.3: Result of *useDifferential* Set to False.

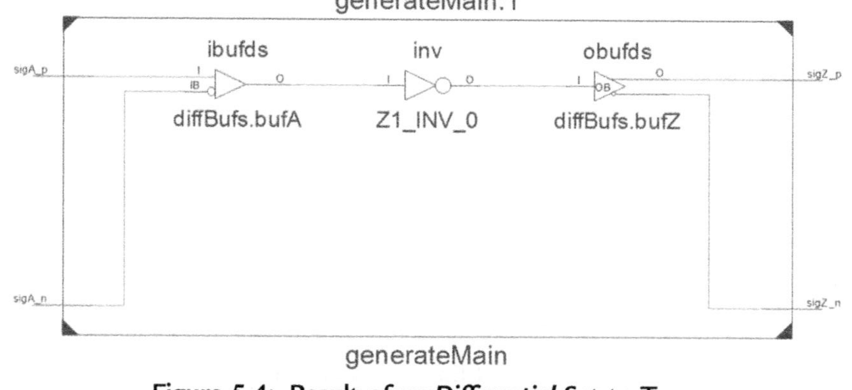

Figure 5.4: Result of *useDifferential* Set to True.

In this example the generic indicates the type of implementation for a specific function or structure. The points to glean from this example:

1. A generic is used to select the type of buffer (or any component). This generic may be selected within the development environment or command line so that the source code need not be modified.
2. The generic has a default value so if this component is instantiated and the generic is omitted by the "calling" component then there will be a known value for the generic.
3. Multiple conditional generates can use a single generic.

Typical uses of this conditional generate is to produce code for various architectures, testing different algorithms, and enabling/disabling diagnostic reporting.

5.3.2 Generate Loop Form

The looping form of the generate statement produces zero or more copies of components or concurrent assignments within the clause.

The typical uses are for instantiating multiple copies of components or generating wide sequential logic.

In its general form:

```
<label>: for <identifier> in <range> generate
  { <concurrent statements>; }
end generate [<label>];
```

Snippet 5.13

where

<label> represents a legal and unique label identifier
<range> is an expression specifying a range such as 0 to 12 or 19 downto 13
<concurrent statements> zero or more legal concurrent statements

Imagine that you are tasked with writing the interface to a 72-bit wide memory device. You could instantiate each bi-directional buffer manually…

```
. . .
mb0: IOBUF port map (I=>outbound_data(0), O=>inbound_data(0),
                     T=>direction, IO=>data_bus_pin(0));
db0: IOBUF port map (I=>outbound_data(1), O=>inbound_data(1),
                     T=>direction, IO=>data_bus_pin(1));
<68 iterations later…>
db0: IOBUF port map (I=>outbound_data(71), O=>inbound_data(71),
                     T=>direction, IO=>data_bus_pin(71));
```

Snippet 5.14: Manual Enumeration of Bus Interface.

Certainly "cut-and-paste" is a possibility, but is tedious and prone to certain types of text entry errors. Manual component instantiation also consumes many lines of code and degrades the code's readability.

OR, you could place a single instance of the bi-directional buffer within a generate loop…

```
. . .
make_data_bus: for i in 0 to 71 generate
      dbx: IOBUF port map (
                          I => outbound_data(i),
                          O => inbound_data(i),
                          T => direction,
                          IO=> data_bus_pin(i));
   end generate make_data_bus;
```

Snippet 5.15: Generate Version of Bus Interface.

Notice the use of the index "*i*" which differentiates one instance from another. This identifier doesn't need to be defined explicitly – it is automatically defined in this context and only valid within this generate loop.

Generate loops may also be used to combine multiple inputs for a wide combinatorial logic function. Assume that a 128-bit wide OR gate must be generated (perhaps to identify if an interrupt occurred in one or more peripherals), then, instead of writing out one painfully long equation, one could simply write:

```
1: intermediate(0) <= peripheral_interrupt(0) OR
peripheral_interrupt(1);
2: intrDetect: for i in 2 to 127 generate
3:     intermediate(i-1) <= intermediate(i-2) OR
                           peripheral_interrupt(i);
4: end generate intrDetect;
5: genIntr <= intermediate(126);
```

Snippet 5.16

where *intermediate* is a std_logic_vector(126 downto 0) and genIntr is the final result.

Code review:

Line 1: gets the ball rolling by producing the first *intermediate* result from the first two *peripheral_interrupts*

Line 2: loops through the remaining *peripheral_interrupt* sources. Notice that *i* is intrinsically defined as an integer. The lower index is set to 2 as the previous line handled the first two interrupts.

Line 3: since the first *intermediate* value was built of the first two signals, intermediate($i-2$) represents the last intermediate value generated when entering this loop. The result is placed in intermediate($i-1$) which is, essentially, the next intermediate position.

Line 5: the last intermediate value in the vector contains the desired result.

Be certain to check the output of your synthesis engine to see how this code is actually implemented. Depending on the architecture and the efficiency of the synthesizer, it may make sense to code this in a different fashion to align more closely with the underlying logic structure of the FPGA you are using.

Notice that this is the literal interpretation of the code created by the generate loop (Figure 5.5 shows the first three levels of logic – the real design would be 127 levels deep!). While this produces the theoretically correct logic, one would hope that this wouldn't be the implementation in the FPGA logic!

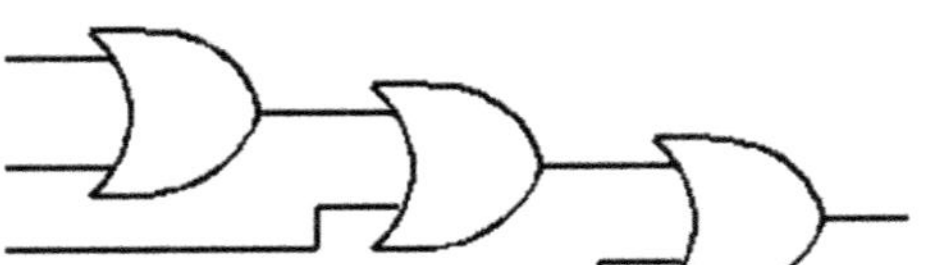

Figure 5.5: Unoptimized Logic Resulting from Generate Loop.

Performance in an FPGA is generally determined by how quickly a signal can move from the output of one Flop to the input of another. Anything between these two flops generally reduces performance. This could be the physical connection between the two devices or it could be a combination of Look-Up Tables (LUTs) and routing resources. If the cascaded OR design from above were implemented as shown in the "Unoptimized Logic" diagram, then 127 "levels of logic" or LUTs would be required which would not only consume a large number of resources (127 LUTs), but also implement extremely slowly.

Recall from the earlier discussion about architecture that most FPGAs use small LUTs to implement logic. Older parts use four input LUTs, while newer LUTs can manage six simultaneous inputs to generate a single output.

Fortunately, the synthesis tools will recognize this as an inefficient design and construct the 128 input OR gate closer to the design shown below for an architecture that uses six input LUTs...

The first level of logic (left side) combines every six inputs into a single OR gate (see Figures 5.6 and 5.7). One hundred and twenty-eight inputs require 22 LUTs (with a few inputs not used). The second level of logic collects every six inputs from the first level of logic using four more LUTs (again, with a few left over). Finally, one more LUT is required to combine the second level of logic into a single-bit result.

Let's do a quick review of Generate and Generic statements:

1. Generics can be of any legal type or subtype.
2. Generics are used in the same way as constants are within a module.
3. Generic values can change from instantiation to instantiation, but are constant within a module.
4. Top-level·Generics can usually be set via synthesis tool options (GUI) or command line arguments so that source code doesn't need to be modified.
5. Generate statements come in two flavors: conditional and looping.
6. Generate statements can wrap one or more concurrent statements – this includes component instantiations, assignments, conditional assignments, and even processes.
7. Generic and Generate statements are often, but not always, used together to make code more flexible for reuse.

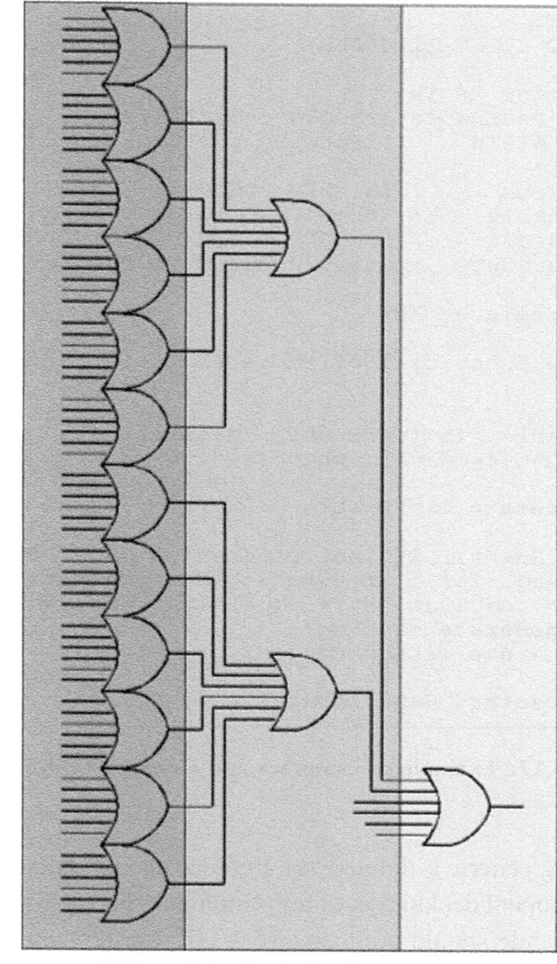

Figure 5.6: Optimized OR Gates.

The score?

	Unoptimized	Optimized
Number of LUTs	127	27
Levels of Logic	126	3

Figure 5.7: Comparison Between Optimized and Unoptimized Synthesis.

Since we're here, let's build an arbitrary width register in which we can tie the use of *generics* with *generates*.

```
 1: library IEEE;
 2: use IEEE.STD_LOGIC_1164.ALL;
 3:
 4: entity register_XE is
 5:    generic (useEnable : boolean := false;
 6:             WIDTH     : integer := 1024
 7:            );
 8:    port ( clock      : in  STD_LOGIC;
 9:           enable     : in  STD_LOGIC;
10:           dataIn     : in  STD_LOGIC_VECTOR (WIDTH-1 downto 0);
11:           dataOut    : out STD_LOGIC_VECTOR (WIDTH-1 downto 0)
12:         );
13: end entity register_XE;
14:
15: architecture Behavioral of register_XE is
16:   begin
17:
18:      withEnable: if (useEnable) generate
19:        dataOut <= dataIn when (rising_edge(clock) and
                                   (enable='1'));
20:      end generate withEnable;
21:
22:      withoutEnable: if (not useEnable) generate
23:        eachReg: for i in 0 to WIDTH-1 generate
24:          reg: dataOut(i) <= dataIn(i) when rising_edge(clock);
25:        end generate eachReg;
26:      end generate withoutEnable;
27:
28:   end architecture Behavioral;
```

Snippet 5.17: Example of Generate and Generics Combined Use.

Code analysis:

Line 5: the *useEnable* generic is defined as a Boolean and defaults to false unless either the instantiation or the component declaration of this module in the "calling" module forces this to "true". If "true" this module should produce a series of registers which condition the clock signal with an enable, otherwise the enable will be omitted.

Line 6: the *width* generic specifies the number of bits that this register can support.

Lines 10,11: Notice the use of the generic *WIDTH* in the range specifier for the std_logic_vectors.

Line 18: if the *useEnable* generic is set to true then everything between this line and the close of the generate on line 20 is evaluated.

Line 19: should the *useEnable* generic be set, then this line of code will generate a flip-flop with a synchronous clock enable for every position within signal *dataIn.*

Line 22: if the *useEnable* generic is set to false then all lines between this one and the end generate on line 26 will be evaluated.

Line 23: this line creates a loop which will execute *WIDTH* times generating anything between this line and the loop close (on line 25). Notice the use of both types of generates and the case where one generate is nested in another.

Line 24: for each value of *i* a single register, without enable, is created. Admittedly, this line implementation is not as "clean" as the equivalent statement on line 19 (minus the enable, that is), and it is coded this way to demonstrate that VHDL often offers multiple ways of producing the same effect.

What gets generated…

Based on the values of the generics, the synthesis tools will produce different logic (see Figures 5.8 and 5.9).

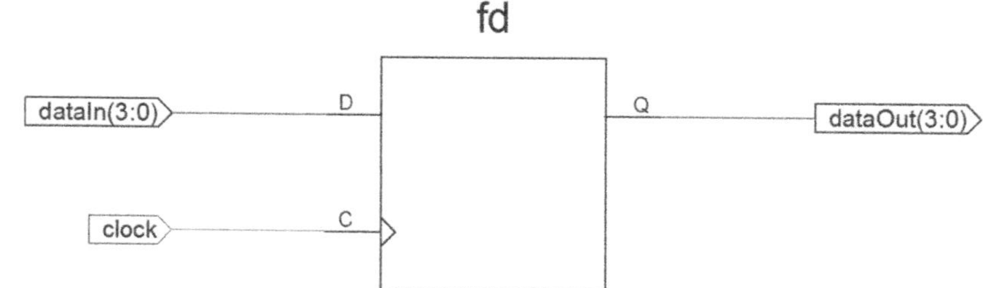

Figure 5.8: Result of Above Example with *useEnable* Set to False and *WIDTH* Set to 4.

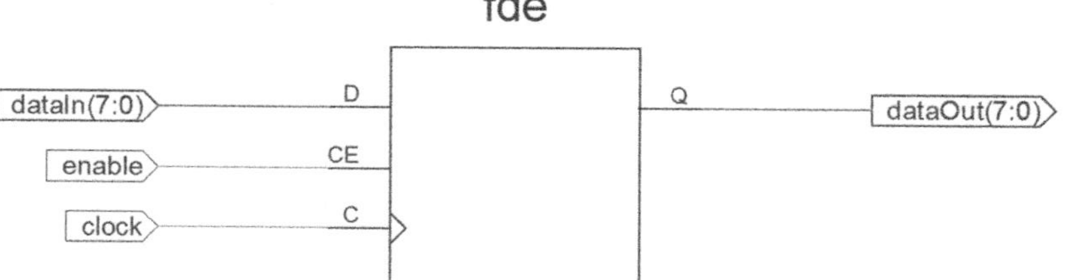

Figure 5.9: Result of Above Example with *useEnable* Set to True and *WIDTH* Set to 8.

Driving the Generics…

Let's now look at how generics are "driven" to complete our understanding of this capability.

Generics will show up in three places in our code. First, it can appear in the module (within the entity statement) as shown in Snippet 5.17. Next, since every module needs to be declared before it can be used, we see the same generic in the component instantiation. Finally, the generic can appear in the actual instantiation.

Wow, generics in three different places – how does one keep track of all of this? The rule is easy – the more "specific" the use is the value that it takes. If the generic is given a default value in the entity of the module, then that is the default value for that generic and will be overridden if the generic is not associated with a value anywhere else.

Next comes the component definition – since component definitions can be placed in other components it is more specific and therefore any value assigned to the generic will take a higher priority than any value given to the generic in the module's entity statement and therefore be used provided that that generic is not assigned a value in the final possible place – in the instantiation.

The instantiation of the module has the highest priority – whatever value is assigned in the instantiation for a given generic is used.

Let's see this in action…

Code analysis:

Lines 4,5: these lines are used to access Xilinx's FPGA primitives.

Lines 18–25: the register_XE component is defined.

Lines 31–38: An enable generator is coded to provide a dynamic enable signal to the registers. This process produces 1 enable for every 4 clock (that is, for every 4 clock cycles, one of those cycles is an enable). This technique implements a circular shift register preloaded with the pattern in which to generate an enable. The leftmost position in the shift register is used as the enable. The advantage to this design is that it uses only one silicon resource – a shift register, and since only one silicon resource is used, it can run at the full speed of the FPGA – this is NOT a performance-limiting design. A counter, on the other hand, would likely require several LUTs and flops and operate at a much lower performance. This technique can be effectively used in the Xilinx Spartan6 and Virtex6 architectures up to 32 bits wide. After that point, more than one shift register will be required and a resource vs. performance analysis would need to be done to determine where the tradeoff point is (for large counters, above about 1024 counts, you may wish to consider the use of Xilinx's DSP48e which can implement a 48-bit counter at full FPGA performance – but there are far fewer of these resources than there are of shift registers and flops).

Lines 40–46: the register_XE component is instantiated without an enable signal and forcing the number of bits for both the *dataIn* and *dataOut* ports to four positions in width.

Lines 48–54: another register_XE component instantiation, but this time using the enable signal and supporting a width of eight positions.

Lines 56–59: These lines instantiate the Xilinx Flip-Flop with Enable primitive (actually a macro) which produces the same effect as the wEnB instantiation. It is provided here as another way to solve the same problem.

```
library IEEE;
use IEEE.STD_LOGIC_1164.ALL;

library UNISIM;
use UNISIM.VComponents.all;

entity makeRegisters is
  generic(WIDTH : integer := 16);
  port (clock         : in  STD_LOGIC;
        dataIn        : in  STD_LOGIC_VECTOR (WIDTH-1 downto 0);
        dataOut       : out STD_LOGIC_VECTOR ( 3 downto 0);
        dataOut2      : out STD_LOGIC_VECTOR ( 7 downto 0);
     dataOut3 : out STD_LOGIC_VECTOR ( 7 downto 0));
end entity makeRegisters;

architecture Behavioral of makeRegisters is

    component register_XE
      generic (useEnable : boolean := false;
               WIDTH     : integer := 4096);
      port(clock   : in std_logic;
           enable  : in  std_logic;
           dataIn  : in  std_logic_vector(WIDTH-1 downto 0);
           dataOut : out std_logic_vector(WIDTH-1 downto 0));
    end component register_XE;

    signal enable : std_logic := 'U';

 begin

    mkEn: process (clock)
       variable count : std_logic_vector(3 downto 0)  := X"1";
      begin
       syncEvents: if rising_edge(clock) then
          enable <= count(3);
          count  := count(2 downto 0) & count(3);
       end if syncEvents;
     end process mkEn;

   noEnA: register_XE
       generic map (useEnable => false,
                    WIDTH     => 4 )
       port map(clock   => clock,
                enable  => enable,
                dataIn  => dataIn(3 downto 0),
                dataOut => dataOut );

   wEnB: register_XE
       generic map (useEnable => true,
                    WIDTH     => 8 )
       port map(clock   => clock,
                enable  => enable,
                dataIn  => dataIn(7 downto 0),
                dataOut => dataOut2 );

   prim: for i in 0 to 7 generate
     flop: FDE port map (D=>dataIn(i), C=>clock, CE=>enable,
                         Q=>dataOut3(i));
   end generate prim;

 end architecture Behavioral;
```

Snippet 5.18: Making Registers.

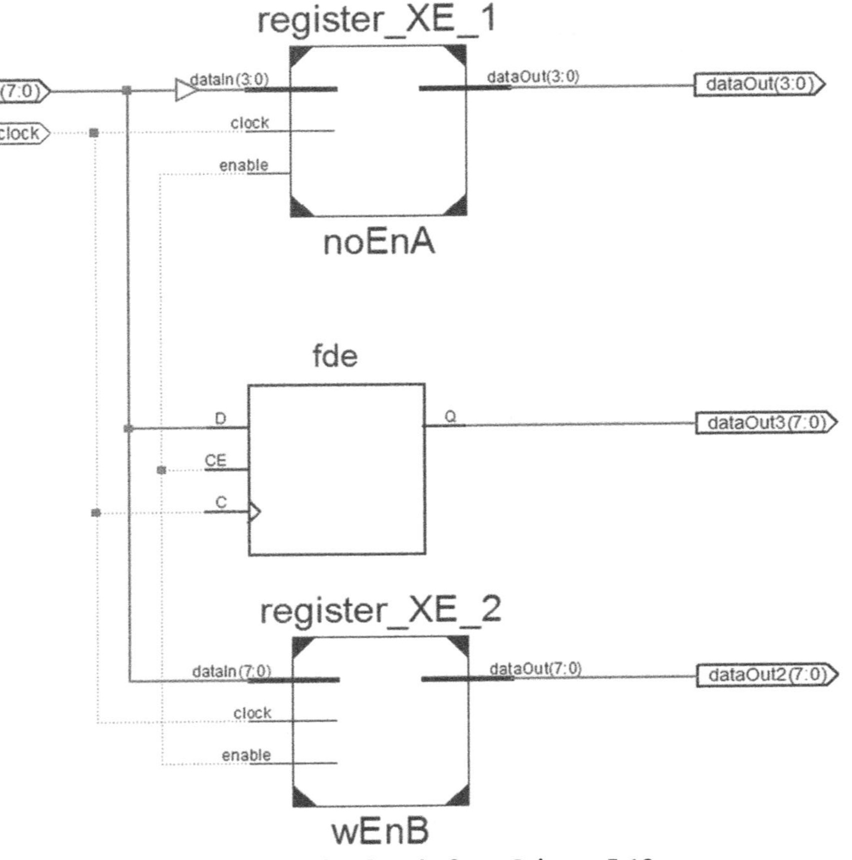

Figure 5.10: Result of Code from Snippet 5.13.

This code produces the result shown in Figure 5.10.

Note: the circuitry for driving the enable has been omitted for clarity.

What if...

...the *useEnable* generic was omitted on line 49?

- Then the value of *useEnable* in the component definition in this module would be used (*useEnable* would be set to false).

...the *WIDTH* generic was omitted on line 42?

- Then the value of the *WIDTH* generic in the component definition in this module would be used (*WIDTH* = 4096). This would also have the effect of creating a width mismatch between the port named *dataIn* and the signal slice *dataIn* (line 45) and between the port named *dataOut* and the signal *dataOut* (line 46) which is defined in the port statement of this module to be eight positions wide, resulting in an error.

...the *WIDTH* generic was omitted both on line 20 and on line 50?

Then the value for *WIDTH* specified in the module containing *register_XE* would be used (line 6 of Snippet 5.13). If *WIDTH* were defined, but had no initial value, then the tools would generate an error.

5.4 Functions and Procedures

Functions and Procedures in hardware are remarkably similar in concept to functions and procedures in software. In software (and hardware simulation), functions and procedures are used to reduce the amount of code generated thus saving memory space, but at the cost of time (due to the overhead of "calling" the function or procedure), whereas, functions and procedures in FPGAs tend to generate hardware which has no time penalty, but may require additional resources.

Both functions and procedures can be found in package bodies, processes (between the process statement and the processes' begin statement), and architectures (between the architecture statement and the architecture's begin statement).

Procedures are blocks of codes which take zero or more "arguments" and perform a designated task. These tasks may change zero or more of the passed "arguments". A procedure does not "return" a value as a function does, but rather relies on modification of one or more of the passed arguments.

Functions are blocks of codes which take zero or more "arguments" and perform a designated task. While these tasks may change zero or more of the "arguments", they must return exactly one value of the designated type.

```
[pure | impure] function <identifier> [(parameter list)]
                              return <type> is [<declarations>]
   begin
      <sequential statements>
   end [function] [<identifier>];
```

Snippet 5.19: Generic Form of a Function.

where

<identifier> is any legal and unique identifier name

<parameter list> is a comma separated list of signals, variable, or constant types

<type> is the type of the value being returned to the caller

<declarations> is zero or more declarative statements which may include variables, constants, types, subtypes, and other functions or procedures. Signals may not be declared within a function

<sequential statements> is a number of sequentially executable statements

There should be at least one occurrence of a return statement in the <sequential statements>.

```
[<label>:] return <expression>;
```

Snippet 5.20: The Generic Return Statement.

Note that the <expression> must be of the same type as is listed in the declaration of the function. Multiple return statements are legal within the body of a function.

The overall structure of Functions and Procedures is similar. Both begin with a statement indicating if it is a function or procedure, what arguments they take, and in the case of the function, the type of value that is returned.

```
[pure | impure] procedure <identifier> [(argument list)] is
     [<declarations>]
   begin
      <sequential statements>
   end [function] [<identifier>];
```

Snippet 5.21: Generic Form of a Procedure.

Both the Function and Procedure contain an area for specifying types, subtypes, subordinate functions and procedures, constants, and variables (but not signals). Whatever types, constants, or variables that are defined in this function or procedure are only available within that function or procedure. Information can only be made available from within the function or package by means of the arguments or the return value in the case of functions.

Next comes the keyword "begin" which marks the beginning of the description of what the function or procedure does. Finally, the keyword "end" with the optional keyword of "function" or "procedure" and, again optionally, the name of the function or procedure.

Functions and Procedures can access all types, subtypes, functions, procedures, constants, generics, signals, and variables of the architecture, process, or package that they are defined in. This is different than the way that the architecture behaves. While the architecture relies on the entity statement to make "connections" to other hierarchical constructs, the function and procedure have access to anything above their current level of hierarchy.

For example, if signal x is defined in an architecture and a function is defined in the same architecture and further defines a variable y, then signal x from the architecture is available in the function, but variable y is only available within the function and not to the other objects in the architecture.

There is a special category of functions and procedures known as "pure" and "impure". Functions are "pure" by default, that is, if you neglect to specify that the function or procedure is pure or not, VHDL assumes that the function or procedure is pure. Pure, in this context, means that all the information that flows into the function arrives via the arguments, and all the information that flows out of the function are returned via the return statement and/or the arguments.

Conversely, an "impure" function or procedure either modifies a value not listed in the parameter list, or uses a value not listed in the parameter list.

```
1: function sl2char (x : std_logic) return character is
2:    begin
3:       case (x) is
4:          when '0'=> return '0';
5:          when '1'=> return '1';
6:          when 'U'=> return 'U';
7:          when others => return '?';
8:       end case;
9:    end function sl2char;
```

Snippet 5.22: Example of a "Pure" Function Taking a Single Parameter.

Code analysis:

Line 1: the function *sl2char* (named for std_logic to character) takes a signal, variable, or constant std_logic as an argument and indicates that it will return a single character.

Lines 4–6: the case statement compares the value of *x* to a '0', '1', or 'U'. If it is any of these three values, then the character representation is returned to the calling routine.

Line 7: since there are nine possible values for a std_logic type, the other possibilities are handled in the "when others" clause, in which case a question mark is returned.

This function is "pure" as it makes no changes to anything outside the function.

"Impure" must be specified for functions that change variables or signals outside the function that are not specified in the argument list. This might produce a situation where the values returned by the function differ even when the parameter list values are the same.

One of the rationales for forcing the designation of "impure" on a function is that this coding style may produce undesirable "side-effect" and change the behavior of the code in unexpected ways.

Let's start with a simple function that "simply" increments a variable:

```
 1: doIncr: process
 2:    variable iVar : integer := 0;
 3:    impure function incr return integer is
 4:       begin
 5:          iVar := iVar + 1;
 6:          return iVar;
 7:       end function incr;
 8:    begin
 9:       iA <= incr;
10:       wait for 1 ns;
11:    end process doIncr;
```

Snippet 5.23: Example of an Impure Function.

Code analysis:

Line 2: variable *iVar* defined as an integer and initialized to zero.

Line 3: impure function *incr* is defined as a function which takes no parameters and returns an integer. Notice that this function is defined within the process statement block.

Line 5: variable *iVar* is incremented. Since there is no *iVar* defined in the function, the scope steps back and looks for the next level in which to check for an *iVar* which is the process level. This means that the *iVar* OUTSIDE the function is going to be incremented, hence the IMPURE label prior to the function.

Line 6: *iVar* is returned to the caller. Notice that this is an integer, just as indicated on line 3.

Line 9: the function *incr* is "called" with no parameters and its value is assigned to signal *iA* (which we can assume has been defined as an integer).

Line 10: the process enters suspension for 1 ns and repeats.

This example builds a counter which counts from 0 to 2^32 − 1 once every nanosecond.

5.4.1 Function and Procedure Parameters

Both functions and procedures can take parameters. These parameters help insulate the effects of a function or procedure from the calling routine. Parameters, like ports in an entity, can be inputs (in), outputs (out), or bi-directional (inout).

Inputs can be "read" or used on the right-hand side of an assignment statement, or in the conditional of a case, while, for, or if statement. Inputs are treated as constants within the function or process.

Outputs can only be "written" to or used as the target of an assignment. If a value is to be used both as a target and as a data source, it can be either defined as a bi-directional signal or a temporary variable can be used for the computation and then assigned to the output.

Each parameter must also have a type, just like a signal in a port description.

Finally, each parameter must have a class. A class can be a constant, variable, signal, or file. We'll skip the discussion of files here as they are generally not used in FPGA design.

All parameters default to a class of "constant" and "in" if not otherwise defined. The constant class forces the parameter to behave as a constant within the function or procedure, that is, it can be used on the right-hand side of an equation or within a conditional test, but cannot have a value assigned to it. Since any parameter designated as an "in" can't change its value, it is, by definition of the "constant" class. Therefore, you don't need to specify both "constant" and "in", "in" takes care of both aspects.

All parameters defined as "out" or "inout" default to a parameter class of "variable".

If a parameter is of the signal class, then that parameter behaves as a signal and assignments made to it must use the signal assignment "<=" notation. This parameter will only take on its new value when the function or parameter is exited. By contrast, if the parameter is defined as a variable it must use the variable assignment ":=" notation and takes on its value immediately.

Additionally, if the parameter is classed as a signal, then the parameter in the calling routine must also be a signal. Similarly if the parameter is classed as a variable, then the parameter in the calling routine must also be a variable.

Recall that both functions and procedures can be called both from within a process or as a concurrent statement, therefore the specification of the parameter as a signal or a variable is important. Of course, you may choose to omit the parameter class in which case the parameter may be either.

We'll look at some more functions and procedures when we cover attributes in a couple of sections.

5.4.2 Overloading

Overloading is a concept that can be applied to both functions and procedures. Simply put, overloading allows for the multiple instances of functions and procedures with the same name, but differing number or types in the parameter list. The synthesis and simulation tools can resolve these differences into unique choices; however, if two or more functions or procedures have the same name, the same order and type of parameters, and in the case of functions, returns the same type, then there is ambiguity as to which function is actually being referred to and an error will be raised.

The value of overloading can be seen with the simple addition function ("+"). Since the addition symbol is defined for the integer type, how is it that we can add time or real types? What happens if you create a new type, let's say color and you want to be able to add colors?

Let's look at one way of doing this. This is neither a complete example, nor is this the best or most accurate way to solve this problem. It does, however, demonstrate the overloading concept.

Code analysis:

Line 1: a user-defined type called *colors* is defined. A small set of names is defined for this new type.

Line 3: the function "+" is defined which takes two parameters, both of the user-defined type *color* and will return a value of type *colors.*

Line 5: the first parameter is analyzed – if the first parameter is *RED, BLUE, YELLOW, WHITE,* or *BLACK*, then the specific case for that color will be handled, otherwise the "when OTHERS" clause on line 39 will be analyzed and the color *BLACK* will be returned. The second

```
type colors is (RED, BLUE, YELLOW, GREEN, ORANGE, PURPLE,
                PINK, WHITE, LIGHT_BLUE, BLACK, GRAY);

function "+" (color1, color2 : colors) return colors is
   begin
      firstColor: case color1 is
         when RED =>
            case color2 is
               when RED    => return RED;
               when BLUE   => return PURPLE;
               when YELLOW => return ORANGE;
               when WHITE  => return PINK;
               when OTHERS => return BLACK;
            end case;
         when BLUE =>
            case color2 is
               when RED    => return PURPLE;
               when BLUE   => return BLUE;
               when YELLOW => return GREEN;
               when WHITE  => return LIGHT_BLUE;
               when OTHERS => return BLACK;
            end case;
         when YELLOW =>
            case color2 is
               when RED    => return ORANGE;
               when BLUE   => return GREEN;
               when YELLOW => return YELLOW;
               when OTHERS => return BLACK;
            end case;
         when WHITE =>
            case color2 is
               when RED    => return PINK;
               when BLUE   => return LIGHT_BLUE;
               when WHITE  => return WHITE;
               when BLACK  => return GRAY;
               when OTHERS => return BLACK;
            end case;
         when BLACK =>
            return BLACK;
         when OTHERS =>
            return BLACK;
      end case firstColor;
   end function "+";
```

Snippet 5.24: Overloading Function for Addition.

level of case statements produces the solution of the two parameters being added. Note that the color additions are not exhaustive and that there are color combinations that are not defined.

5.4.3 When to Use Procedures and Functions

As a general rule, functions are used to modify a particular piece of data. For example, computing the next value of a gray code sequence can easily be computed with a function. The function would take the current gray code value as an argument (input only), compute the next value, typically as a std logic_vector, then return that new value. No values outside the function are modified indicating that this could be implemented as a pure function.

Procedures are used when several values must be changed or when no values are returned. Examples of use might be a procedure for initializing a large set of data, or writing specific messages to the simulation console.

Do functions or procedures generate code? Sometimes! Although many designers prefer to use functions and procedures within the simulation environment (where no hardware is generated), both functions and procedures are synthesizable constructs. In fact, most of the IEEE library packages are comprised mainly of synthesizable functions.

Even though a function can be synthesizable, it doesn't mean that it will necessarily generate logic. As an example, a simple function can be written to reverse the bit ordering in a bus:

```
1: pure function reverse (x : std_logic_vector)
                                        return std_logic_vector is
2:       variable slvToReturn : std_logic_vector(31 downto 0) :=
                                                      (others=>'U');
3:    begin
4:       swap: for i in 0 to 31 loop
5:          slvToReturn(i) := x(31 - i);
6:       end loop swap;
7:       return slvToReturn;
8:    end function reverse;
```

Snippet 5.25: Function Demonstrating Non-logic Producing Code.

Code analysis:

Line 1: function *reverse* is explicitly defined as a pure function which takes a std_logic_vector and returns a std_logic_vector. Notice that ranges are not specified.

Line 2: a temporary variable named *slvToReturn* is defined as a 32-bit std_logic_vector. This will require that all values of *x* will need to be 32 bits. This limits the flexibility of this function. We will revisit this in a couple of sections to demonstrate how the use of Predefined Attributes can improve reusability.

Line 4: each position within *x* is enumerated. Notice again that this limits *x* to exactly 32 bits (starting at index 0).

Line 5: position *i* in *the* temporary variable *slvToReturn* is set to the value of position (31–1) in parameter *x*. This effectively reverses the vector. Remember that although this appears to be a sequential statement, it is "unrolled" during elaboration to become logic, or in this case, connections.

Clearly, this example illustrates that connections are rearranged, but this module doesn't produce any logic (therefore no LUTs are implemented), nor is any information stored/buffered (therefore no Flops are required) (see Figure 5.11).

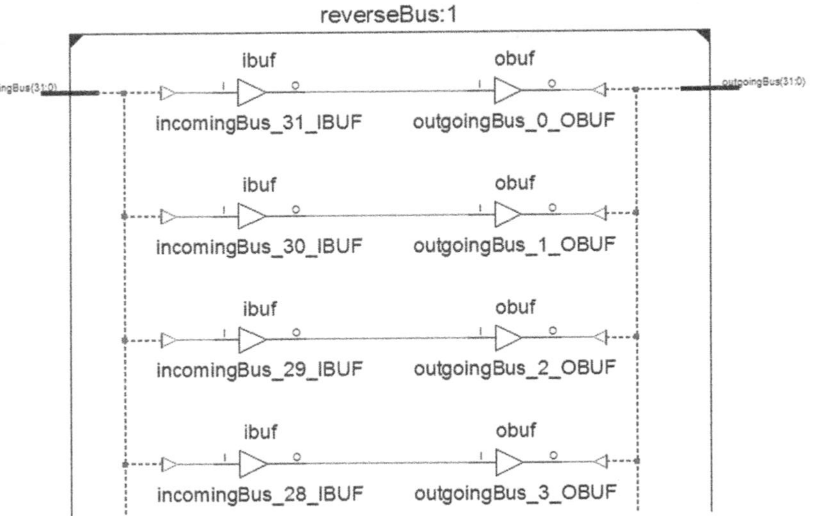

Figure 5.11: Result of Implementation of the Reverse Function.

Note: signals *incomingBus* and *outgoingBus* were attached to the ports in the entity section. The synthesis tools automatically insert buffers going to and coming from ports. Note that incomingBus_31 is connected to outgoingBus_0. The buffers (ibuf and obuf) represent buffers of carrying signals in and out of the FPGA and are not generated by the function.

5.4.4 Using Functions and Procedures

We've spoken about how to create a function and procedure, how to determine if it's pure or impure, aspects of the parameter list, and a couple of examples.

Only two easy pieces remain to begin using functions and procedure. First – where to define a function or procedure, and second, how to invoke the function or procedure.

First things first, VHDL requires everything to be defined before it can be used. We've seen how to define functions and procedures, but where to put these definitions? There are only two places that they may reside: within a declarative block or within a package. Packages are the topic of the next section, so we'll cover how functions and procedures are handled in packages in a moment, so, for now, we'll focus on how function and procedure can be defined in the declarative statements.

Declarative statements are found in a number of locations (see Figure 5.12).

Each location has its own scoping limits. For example, if a function or procedure were defined in the entity statement, then all architectures associated with that entity would have access to that function or procedure.

If the function or procedure were defined in the architecture's declarative area then that function or procedure would be accessible anywhere in that architecture (but not

```
entity ABC is
   <declarative area>
end entity ABC;
```

```
architecture BEH of ABC is
      <declarative area>
   begin
      . . .
   end architecture BEH;
```

```
xyz: process (clk)
      <declarative area>
   begin
      . . .
   end process xyz;
```

```
function f1 (…) return … is
      <declarative area>
   begin
      . . .
   end function f1;
```

```
procedure p1 (…) is
      <declarative area>
   begin
      . . .
   end procedure p1;
```

Figure 5.12: Declarative Areas Available for Function and Procedure Definitions.

available in other architectures that use the same entity as is the case with the definition within the entity).

If the function or procedure were defined in a process, another function, or another procedure, then it would only be accessible within that process, function, or procedure.

Let's say you write a function or procedure that you want to use in multiple architectures, where would you define your function or procedure? This is where packages come in very handy. We'll cover the details of how to create a package in the next section, but for now just know that it is possible to define functions and procedures (and other objects) within a package.

Now this package that contains your special functions and procedures can be accessed by including the "use" statement and naming the library and package that contains the items that you want to use.

So, how do you make the decision as to where to define your function or procedure? Many designers use the principle of reuse to guide their decision. If the function is relatively generic, such as big-to-little endian, bit reversal, simulation "helper" functions such as pseudo-random noise generators, and display utilities, then the function or procedure should probably be put in a non-project specific package as it is likely to be reused in a number of projects.

If the function or procedure is not generic enough to be used by other projects, but is frequently used within a design, then it should be placed in a project specific package.

If a function or procedure is only used within a specific architecture, process, function, or procedure, then it should be defined only within that object. This limits the scope of the function or procedure and may avoid naming issues (since the scope is limited, functions or procedures defined in different processes can have the same names without conflict. This, however, is not always the best practice as it may lead to confusion – always name your function and procedures as you name your identifiers – long enough to be clearly understood).

> **Best Design Practices Tip 5.1 –**
>
> Make every attempt to write functions and procedures that can be used in other projects in a generic fashion and place them in an appropriate package.
> Corollary - If the function or procedure is specific to the project that you are working on and will be used more than once in the project, add it to a project specific package.
> Corollary - If the function or procedure is specific only to the few instances within an architecture, process, function, or procedure, then code the function or procedure in the highest level in which it is used and no higher.

Remember that functions and procedures can be defined in any or all of these locations – they all don't have to go in one place.

Now that we know where to stick functions and procedures, let's finally look at invoking them.

Like component instantiations, there are two basic ways to pass the parameters to both functions and procedures: by positional association and by named association. For exactly the same reasons given favoring named association in component instantiation, named association is favored here as well.

Despite this strong suggestion many designers use positional assignment because functions and procedures typically (but certainly not always) have relatively few parameters.

Let's consider the trivial cases of a function and procedure with no parameters:

```
procedure printTime is
   begin
      report "The simulation time is now: " & time'image(now);
   end procedure printTime;

impure function monitorSigs return std_logic is
   begin
      if (slvValue = X"1234") and (slValue = '1') then
         return '1';
      else
         return 'Z';
      end if;
   end function monitorSigs;
```

Snippet 5.26: Function and Procedure with No Parameters.

Code analysis:

Line 1: procedure *printTime* is defined with no parameters.

Line 3: a simple message indicating the simulation time is pushed to the console. Notice the use of the predefined attribute 'image. "Now" is a predefined variable that exists within the simulation environment. It returns the times since the beginning of the simulation.

Line 6: an impure function *monitorSigs* defined to take no parameters and return a std_logic type.

Lines 8–11: a signal or variable *slvValue* and *slValue* are tested for a specific condition. If the condition is true, then the function returns a '1' otherwise it returns a 'Z'.

And here's how they can be called...

```
doStuff: process
   begin
      printTime;
      wait for 1.5 ns;
   end process doStuff;

doMonitor: process
   begin
      signalResult1 <= monitorSigs;
      wait for 1 ns;
   end process doMonitor;
```

Snippet 5.27: Accessing Procedure *printTime* and Function *monitorSigs*.

We'll assume that *signalResult1* is defined as a std_logic and that *slvValue* and *slValue* are defined and in the scope of function *monitorSigs.*

Let's modify the *monitorSigs* function so that it takes a std_logic_vector and std_logic as parameters...

```
1: function monitorSigs (signal x : std_logic_vector;
                         signal y : std_logic) return std_logic is
2:    begin
3:       if (x = X"1234") and (y = '1') then
4:          return '1';
5:       else
6:          return 'Z';
7:       end if;
8:    end function monitorSigs;
```

Snippet 5.28: Function *monitorSigs* Re-written to Accept Two Parameters.

Code analysis:

Line 1: parameters *x* and *y* are defined as signal class (that is, only signals can be applied to *x* and *y*) and determined to be of type std_logic_vector and std_logic, respectively. This function still returns a std_logic. Notice that this is no longer an impure function as the parameter list defines everything that gets used in this function.

Line 3: this line has been modified to use x and y rather than the signals *slvValue* and *slValue*.

We can use this function using either positional association:

```
signalResult <= monitorSigs(slvValue,slValue);
```

Snippet 5.29: Calling the Modified *monitorSigs* with Position Association.

What happens if we reverse the order of the parameters? With positional association since the types are different, the tools can recognize that parameters aren't lined up properly and will indicate an error.

What happens if the parameters are of the same type? Then it's going to be a long night looking for this problem!

This kind of mix-up is easily avoidable when using named associations:

```
signalResult <= monitorSigs(x=>slvValue,y=>slValue);
```

Snippet 5.30: Calling the Modified *monitorSigs* with Named Association.

Best Design Practices 5.2 – Tip: Using Named and Positional Association

Always use named association when instantiating components or using functions and procedures.
Caveat – Positional association is safe when there is either only a single parameter or each parameter is of a unique type and there is no possibility of confusion.

5.5 Attributes

VHDL possesses a powerful capability known as Predefined Attributes. These Predefined Attributes, sometimes referred to as "tic" attributes due to the use of the single quote (') in the syntax, enhance the ability to "genericize" functions as they allow certain parameters such as the size of an array, the direction of the array index, and other qualities to be discovered and leveraged during synthesis.

Some of these attributes are synthesizable, and we'll focus on some of those here, while other attributes can only be used in the simulation environment. Additionally, not every attribute can be applied to every object.

```
<item>'<attribute name>
```

Snippet 5.31: Syntax for the Predefined Attribute.

where:

<item> – any legal and previously defined type, array, or signal name
<attribute name> – the name of a Predefined Attribute

This syntax is valid for all Predefined Attributes, both synthesizable and not.

The easiest way to learn about the Predefined Attributes is to show a small example which focuses on how these Predefined Attributes can be applied to the commonly used std_logic_vector.

- ‘left – returns the leftmost value in the range
- ‘right – returns the rightmost value in the range
- ‘high – returns the largest value in the range (regardless of order)
- ‘low – returns the smallest value in the range (regardless of order)
- ‘range – returns the range of the array
- ‘reverse_range – the reversed range of an array

Suppose you had two std_logic_vectors – one having an ascending index and the other a descending index as follows:

```
signal up : std_logic_vector(0 to 31);
signal dn : std_logic_vector(31 downto 0);
```

Snippet 5.32

Then…

Table 5.1

Attribute	When Applied to *up*	When Applied to *dn*
‘left	0	31
‘right	31	0
‘low	0	0
‘high	31	31
‘range	0 to 31	31 downto 0
‘reverse_range	31 downto 0	0 to 31
‘ascending	true	false
‘descending	false	true
‘length	32	32

Now, let's look at another example to illustrate how the contents of an array might be accessed. Given the following declarations:

```
type things is (HAT, GLASSES, JACKET, SHIRT, WATCH, GLOVES,
                BELT, KEYS, WALLET, PANTS, SOCKS, SHOES);
subtype sgniht is things range SHOES downto HAT; -- reverses the order
```

Snippet 5.33

- 'pos – returns the index of the specified item within the list
  - <item>'pos(<object>) returns an integer corresponding to where <object> is within the list. The first item in the list is number zero
  - Ex: things'pos(HAT) is 0 – the first item in the list
  - Ex: things'pos(GLOVES) is 5
  - Interesting application: type CHARACTER is an array of characters listed in the ASCII standard order. This enables a very easy conversion from a character to its equivalent ASCII value by calling out
    - CHARACTER'pos('A') which will return 65
- 'val – returns the value at the specified offset into the list
  - <item>'val(index) returns the indexth object in the list
  - Ex: things'val(3) is SHIRT
  - Ex: things'val(11) is SHOES
  - Interesting application: as this is the inverse function to 'pos, if the ASCII value is passed then the equivalent character will be returned
    - CHARACTER'val(65) returns the character 'A'
- 'succ – the next value in the list, regardless of the orientation of the type
  - Ex: things'succ(GLOVES) is BELT
  - Ex: sgniht'succ(GLOVES) is BELT
- 'pred – the previous value in the list regardless of the orientation of the type
  - Ex: things'pred(GLOVES) is WATCH
  - Ex: sgniht'pred(GLOVES) is WATCH
- 'rightof – the item to the right of the specified item within the list dependent on orientation
  - Ex: things'rightof(GLOVES) is BELT
  - Ex: sgniht'rightof(GLOVES) is WATCH
- 'leftof – the item to the left of the specified item within the list
  - Ex: things'leftof(GLOVES) is WATCH
  - Ex: sgniht'leftof(GLOVES) is BELT

Let's cover one more pair which is often used in debugging to write values out to the simulation console.

- 'image – returns a string representing the parameter
  - Examples:

  - integer'image(123) returns the string "123"
  - boolean'image(bX) returns either "true" or "false"
  - character'image('x') returns the string "x"
  - real'image(123.456) returns the string 1.234560e+02
  - std_logic'image('U') returns the string "U"

- Even for user-defined types:
  - things'image(SOCK) returns the string "sock"
- 'value – returns a type built from the passed string (limited capabilities)
  - Examples:
    - boolean'value("true") evaluates to the Boolean value true
    - integer'value("123") evaluates to an integer of value 123

It is important to realize that 'image and 'value may not always be synthesizable or available for every type created. Please refer to the user guide for your particular synthesis/simulation tools.

Since many implementations don't support std_logic_vector'image(), let's write the code and add it to a package. First let's look at the function that takes in a std_logic_vector and returns a string corresponding to that std_logic_vector (SLV).

```
1: function SLV_to_String(x : std_logic_vector) return String is
2:    variable returnString : string(1 to x'length);
3:  begin
4:   for i in 0 to x'length-1 loop
5:      returnString(i+1) := std_logic'image(x(i))(2);
6:    end loop;
7:   return returnString;
8:  end function SLV_to_String;
```

Snippet 5.34

Critical line analysis:

Line 1: declaration of the function – accepts a std_logic_vector and returns a string

Line 2: declaration of the variable to hold the constructed string. Since strings are not dynamically adjustable in VHDL, the string must be sized only once. The final string will be the same size as the length of the std_logic_vector. Notice the use of the 'length attribute

Line 4: loop structure to range from 0 to 1 minus the length to align with the std_logic_vector

Line 5: std_logic'image(x(i)) is used to convert one position at a time to a string. The string that is returned is three characters long with the first and the last being double-quotes. This leaves the actual value of interest right in the middle – at position 2. Since strings default to beginning at offset 1, not zero, returnString adds this offset

Adding this function to a package is straightforward. The function declaration goes in the package, while the entire function goes in the function body. Recall that the package describes only the format of the functions and procedures in the package body that are visible to the user, not the functionality.

```
function <function_name>  ([signal] <identifier_name> :
                                 <direction><type_declaration>
                        { [,[signal] <signal_name> :
                                 <direction><type_declaration>]}
                           )  return <type_declaration>;
```

Snippet 5.35: Generic Description for a Function Header.

where:

<function_name> is the name of the function and follows the standard naming conventions

[signal] is the optional keyword which indicates that the following identifier is a signal rather than the default type of variable

<direction> represents if the previously mentioned identifier name is an input, output, or bi-directional to this function

<type_declaration> is the type of the <identifier_name>. The final <type_declaration> is the type that is returned by this function.

5.6 Packages

Since we've been talking about packages in a general fashion, let's look at how to construct packages and include them with libraries...

Packages are a convenient way to group function definitions, procedure definitions, types, subtypes, constants, and component definitions together (see Figures 5.13 and 5.14). Packages themselves are then gathered together to form libraries.

We've already seen the use of the IEEE library and a number of its functions (see Figure 5.15).

The package is a primary construct and usually, but not always, is accompanied by the package body, a secondary construct. Packages hold "passive" information – "shapes" of things, but don't do any real work. The Package Body contains the behavioral description of any functions or packages described in the Package.

The package and the package body may be in different files, although many designers keep the two sections together (see Figure 5.16).

```
library IEEE;
use IEEE.STD_LOGIC_1164.all;

package example_pkg is

  function SLV_to_String(x : std_logic_vector) return String;

end package example_pkg;

package body example_pkg is

  function SLV_to_String(x : std_logic_vector) return String is
      variable returnString : string(1 to x'length);
    begin
      for i in 0 to x'length-1 loop
        returnString(i+1) := std_logic'image(x(i))(2);
      end loop;
      return returnString;
  end function SLV_to_String;

end package body example_pkg;
```

Figure 5.13: Example Package and Package Body.

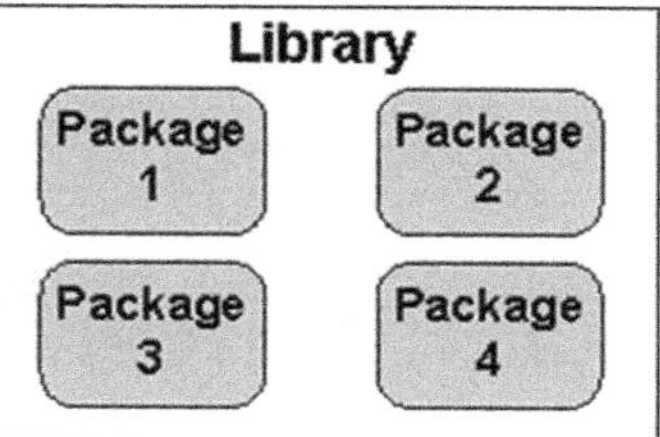

Figure 5.14: Relationship Between Libraries and Packages.

Think of the package as containing the "shapes" of functions – much like a header file may contain function prototype in C. It defines what the function or procedure name is, what kind of parameters it takes, and in the case of a function, what the function returns. The body contains a copy of what was defined in the package except that it adds the functionality to the function or procedure.

The package also acts as a filter – only the functions and procedures that are listed in the package can be accessed by whatever entity or architecture "uses" the package. This means that there can be additional functions or procedures inside the package body that aren't available to the outside world.

Why would someone write functions or procedures that weren't available outside the package body? Assume that there is a fairly complex function or procedure in the package body. The author of the package body might code the function or procedure as a series of "calls" to other functions and/or procedures to make the writing of the code easier. These intermediate functions

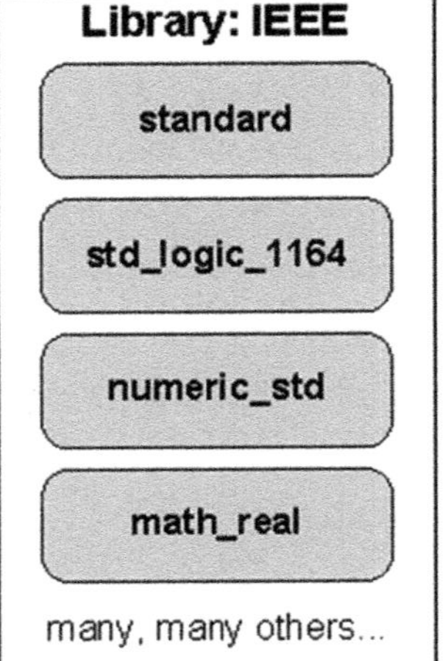

Figure 5.15: Relationship Between the IEEE Library and Some of Its Packages.

```
package ABC is

   function f1 (x : std_logic) return boolean;

   type t1 is (A,B,C);

end package ABC;
```

```
package body ABC is

   function f1 (x : std_logic) return boolean is
      begin
         if (x = 'U') then
            return true;
         else
            return false;
         end if;
      end function f1;

end package body ABC;
```

Figure 5.16: Relationship Between Package and Package Body.

and procedures are generally of no interest to the designer at a higher level. The designer who uses this package only needs access to the "top function/procedure" and none of the supporting functions and procedures, so these supporting functions and procedures are present in the package body, but not listed in the package. So now lets take a look at building a package...

```
package <package name>
   <declarative statements>
end package <package name>;
```

Snippet 5.36: Generic Form of a Package.

Where <declarative statements> represents any combination or order of function "prototypes", procedures "prototypes", types, subtypes, constants, signal, alias, attribute, and/or component declarations.[4]

```
package body <package name>
   <functions and/or procedures>
end package body <package name>;
```

Snippet 5.37: Generic Form of a Package Body.

Where <declarative statements> represents any combination or order of function "prototypes", procedures "prototypes", types, subtypes, constants, and/or component declarations.

```
library IEEE;
use IEEE.STD_LOGIC_1164.all;

package utilities_pkg is

   function reverseVector (signal vecIn : in std_logic_vector)
                                  return std_logic_vector;
   function conv_big_little_endian(signal sigIn : in
                  std_logic_vector) return std_logic_vector;

   signal vcc : std_logic_vector(31 downto 0) := (others=>'1');
   signal gnd : std_logic_vector(31 downto 0) := (others=>'0');

end package utilities_pkg;

package body utilities_pkg is

  function reverseVector (signal vecIn : in std_logic_vector)
                                  return std_logic_vector is
     variable result: std_logic_vector(vecIn'LENGTH-1
                                             downto 0);
  begin
     for i in result'RANGE loop
        result(vecIn'LENGTH-i-result'LOW-1) := vecIn(i);
     end loop;
     return result;
  end function reverseVector;

  function conv_big_little_endian(signal sigIn : in
           std_logic_vector) return std_logic_vector is
     variable output : std_logic_vector(sigIn'LENGTH-1
                                             downto 0);
  begin
     for i in 0 to (sigIn'Length-1)/8 loop
        output := output((sigIn'Length-8-1) downto 0) &
                              sigIn((i*8+7) downto (i*8));
     end loop;
     return(output);
  end function conv_big_little_endian;

end package body utilities_pkg;
```

Snippet 5.38: Another Example of a Package.

[4] There are other items which are legal inside a package; however, these items are far less frequently used by FPGA designers.

Code analysis:

Line 1: the *IEEE* library is declared so that we can then…

Line 2: have access to all of the *STD_LOGIC_1164* package which contains the definitions for std_logic_vector.

Line 4: marks the beginning of the *reverseVector* function. This function is "pure" (the default) as it only uses the parameters and a local variable and will return a std_logic_vector. Notice the absence of any kind of behavioral description – this is handled in the package body.

Line 6: another pure function is defined, this time *conv_big_little_endian,* taking a single std_logic_vector parameter (which must be a signal) and returning a std_logic_vector/.

Lines 9, 10 – two signals are defined – each a 32-bit std_logic_vector, but *vcc* is initialized to all '1's while *gnd* is initialized to all '0's. These values could (and possibly should) be defined as constants, however, some synthesis tools won't accept constants as inputs into port maps when instantiating a component. One should take care not to change these values.

Line 12 – the package ends.

Line 14 – the package body begins. Note that the package body associates with the package through the use of the same package/package body name.

Line 16 – function *reverseVector* is defined, this time with the keyword "is", which begins the description of the behavior for this function.

Line 17 – variable *result* is defined. Notice the use of the attribute to automatically adjust for any length of *vecIn.*

Line 19 – beginning of the loop structure which iterates through each member of *vecIn* signal. One key point to note here is that if the range of *vecIn* is not required to start at 0; however, the definition of *result* does require that the range should start at 0. This is NOT a good coding style, but done here to illustrate how attributes can be used.

Lines 25–33: this function is very similar to the previous function, but with the indexes being handled a bit differently. It is also important to note that *sigIn* should have a length evenly divisible by 8 as this function is a byte-oriented swapper.

Best Design Practice: Typical Uses of Packages in Industry

- Naming convention is to end the name of each package with _pkg so that it is readily recognized as a package.
- Although packages and their corresponding package bodies may be in different files, most designers prefer to keep the package and its associated body in the same file.
- Some designers choose to create a separate package that contains all of the component definitions for a project, in this way if a component changes, the change only needs to be made to the one package and wherever the package is used.
- Packages are physically located in a separate set of directories. Each directory contains the packages for a specific library.

- Packages are usually collected into libraries when:
  - There are many packages and the packages are grouped for logistical reasons by similar function or relation.
  - The package or packages are intended to be used throughout a group (small team through company-wide to public distribution).

5.6.1 Organizing and Creating a Library

Libraries are handled differently from environment to environment. Some tools/environments keep packages in a library collected in separate directories, while other tools/environments keep all files in a flat directory.

Recall that libraries are logical, not physical constructs. As such, when packages are loaded into a Project Navigator project, there is an option to select the library in which the package should be placed. If no library is explicitly selected, the "WORK" library is used by default.

Recall that the "WORK" library is where all the sources are placed and is one of the default libraries that is always available without having to explicitly declare.

Each tool vendor will provide instructions as to how packages can be grouped and assigned to libraries. Typically when files are imported into the environment, an option will be provided to specify which library the file should reside in. In the absence of a library name, the file will usually be placed into the WORK library.

5.7 Commonly Used Libraries

The IEEE libraries are approved in the same fashion as the language itself; however, they are treated separately so that modifications to existing packages and new packages can be added as needed.

This section describes the three most commonly used libraries and the most relevant aspect of each.

5.7.1 Simulation Packages

When performing behavioral simulations, generally no special simulation packages are required. Vendor specific simulation packages are needed when using vendor specific primitives or performing a timing-based simulation (post-map and/or post-place-and-route). These timing packages contain information about how long it takes each resource to rise, fall, transition to/from a high-impedance state, etc. Xilinx primitives for behavioral simulation is located in the UNISIM library in the VComponents package. Xilinx's timing simulation library is automatically inserted when performing a timing simulation.

Generally, timing simulation is quite time-consuming and can better be handled by running static timing analysis which is part of the implementation tool set. It is a rare case when a design behaviorally simulates correctly, is properly constrained, and passes static timing analysis fails to operate in the FPGA.

Your specific FPGA vendor will provide you with the appropriate simulation packages.

5.7.2 IEEE_1164

For the FPGA designer, this is the most important of the packages in the IEEE library. This package contains the definition of the std_logic and std_logic_vector, the system of multi-valued logic consisting of nine possible values for a signal:

Table 5.2

Value	Meaning
'U'	Uninitialized
'X'	Strong unknown logic value
'0'	Strong logic zero
'1'	Strong logic one/high
'Z'	High impedance
'W'	Weak unknown logic value
'L'	Weak logic zero
'H'	Weak logic one
'-'	Don't care

To the FPGA designer, '0', '1', and 'Z' are used in synthesizable code for value assignments and conditional tests. When a high impedance is used on a "top-level" signal, most tools will infer a tri-state I/O buffer. When used at a lower level, this "high-impedance" state is usually converted into equivalent non-tri-state logic for performance as true tri-state logic in FPGA fabric don't scale well in the silicon process (the size of the transistor on the silicon) and take longer to transition in and out of tri-state than just switching between a '0' and a '1'.

The other values can't be used in synthesis (some may be synthesizable, but later optimized out), but prove to be very handy when performing both FPGA and board-level simulations.

Shorthand functions such as rising_edge and falling_edge are included in this package.

Also included in this package are a series of logical functions that can be applied to the std_logic/std_logic_vector type including AND, NAND, OR, NOR, XOR, XNOR, and NOT. Conversion functions to and from std_logic/std_logic_vector are included for various bit types. It is important to note that conversions regarding integers and std_logic_vector types are NOT in this package. They can be found in the NUMERIC_STD package which is covered in the next section.

5.7.3 NUMERIC_STD

The NUMERIC_STD package introduces two new types: that of the UNSIGNED and SIGNED arrays of std_logic. Recall that std_logic_vector is merely a pattern with no positional meaning given to the bits. Both the UNSIGNED and SIGNED data types do assume a positional meaning with the least significant position (slice 0) representing the 1's place.

Basic math functions, comparison functions, and conversion functions are present in this package and outlined in Table 5.3…

Table 5.3

Function	Arg 1	Arg 2	Returns	Description
abs	SIGNED	N/A	SIGNED	Absolute value of arg1
−	SIGNED	N/A	SIGNED	Unary minus
+	UNSIGNED	UNSIGNED	UNSIGNED	Adds two vectors of same or different lengths
+	SIGNED	SIGNED	SIGNED	
+	UNSIGNED	NATURAL	UNSIGNED	
+	NATURAL	UNSIGNED	UNSIGNED	
+	INTEGER	SIGNED	SIGNED	
+	SIGNED	INTEGER	SIGNED	
−	UNSIGNED	UNSIGNED	UNSIGNED	Subtracts two vectors of same or different lengths
−	SIGNED	SIGNED	SIGNED	
−	UNSIGNED	NATURAL	UNSIGNED	
−	NATURAL	UNSIGNED	UNSIGNED	
−	INTEGER	SIGNED	SIGNED	
−	SIGNED	INTEGER	SIGNED	
*	UNSIGNED	UNSIGNED	UNSIGNED	Multiplies two vectors of same or different lengths
*	SIGNED	SIGNED	SIGNED	
*	UNSIGNED	NATURAL	UNSIGNED	
*	NATURAL	UNSIGNED	UNSIGNED	
*	INTEGER	SIGNED	SIGNED	
*	SIGNED	INTEGER	SIGNED	
/	UNSIGNED	UNSIGNED	UNSIGNED	Divides two vectors of same or different lengths
/	SIGNED	SIGNED	SIGNED	
/	UNSIGNED	NATURAL	UNSIGNED	
/	NATURAL	UNSIGNED	UNSIGNED	
/	INTEGER	SIGNED	SIGNED	
/	SIGNED	INTEGER	SIGNED	
rem	UNSIGNED	UNSIGNED	UNSIGNED	Remainder function
rem	SIGNED	SIGNED	SIGNED	
rem	UNSIGNED	NATURAL	UNSIGNED	
rem	NATURAL	UNSIGNED	UNSIGNED	
rem	INTEGER	SIGNED	SIGNED	
rem	SIGNED	INTEGER	SIGNED	

(*continued on next page*)

Table 5.3 *(continued)*

Function	Arg 1	Arg 2	Returns	Description
mod	UNSIGNED	UNSIGNED	UNSIGNED	Modulus function
mod	SIGNED	SIGNED	SIGNED	
mod	UNSIGNED	NATURAL	UNSIGNED	
mod	NATURAL	UNSIGNED	UNSIGNED	
mod	INTEGER	SIGNED	SIGNED	
mod	SIGNED	INTEGER	SIGNED	
>, <, =, /=, <=, =>	UNSIGNED	UNSIGNED	boolean	Comparison functions
>	SIGNED	SIGNED	boolean	
>	UNSIGNED	NATURAL	boolean	
>	NATURAL	UNSIGNED	boolean	
>	INTEGER	SIGNED	boolean	
>	SIGNED	INTEGER	boolean	
shift_left	UNSIGNED	NATURAL	UNSIGNED	Shift left function
shift_left	SIGNED	NATURAL	SIGNED	
shift_right	UNSIGNED	NATURAL	UNSIGNED	Shift right function
shift_right	SIGNED	NATURAL	SIGNED	
rotate_left	UNSIGNED	NATURAL	UNSIGNED	Shift left function
rotate_left	SIGNED	NATURAL	SIGNED	
rotate_right	UNSIGNED	NATURAL	UNSIGNED	Shift right function
rotate_right	SIGNED	NATURAL	SIGNED	
resize	SIGNED	NATURAL	SIGNED	Returns a signed of a new size
resize	UNSIGNED	NATURAL	UNSIGNED	
TO_INTEGER	UNSIGNED	N/A	NATURAL	Converts an unsigned to an unsigned integer
TO_INTEGER	SIGNED	N/A	INTEGER	Converts a signed to an integer
TO_UNSIGNED	NATURAL	NATURAL	UNSIGNED	Converts a non-negative integer to an unsigned of a specified length
TO_UNSIGNED	STD_LOGIC_ VECTOR	N/A	UNSIGNED	Converts an SLV to an unsigned
TO_SIGNED	INTEGER	NATURAL	SIGNED	Converts an integer to a signed of a specified length
TO_SIGNED	STD_LOGIC_ VECTOR	N/A	SIGNED	Converts an SLV to an unsigned
TO_ STDLOGICVECTOR	UNSIGNED	N/A	STD_LOGIC_ VECTOR	Converts an unsigned to an SLV
TO_ STDLOGICVECTOR	SIGNED	N/A	STD_LOGIC_ VECTOR	Converts SIGNED to STD_LOGIC_ VECTOR
not	UNSIGNED	N/A	UNSIGNED	Inverts arg
and, or, nand, nor, xor, xnor	UNSIGNED	UNSIGNED	UNSIGNED	Performs specified logic function
not	SIGNED	N/A	SIGNED	Inverts arg
and, or, nand, nor, xor, xnor	SIGNED	SIGNED	SIGNED	Performs specified logic function

The reason for multiple entries of the same function name in this table is to show which types are supported in overloading.

For example, "NOT" can take either a signed or unsigned argument (in addition to what is supported in the STANDARD and IEEE_1164 packages). Attempting to apply the NOT operator/function to a real number is not defined in this package, nor any other, and will therefore produce an error during synthesis or simulation.

It is important to note that not every synthesis tool will implement multiplication and division for every possible argument. For example, some synthesis tools may only support multiplication by some power-of-two as this is easily done by wire shifting and requires no logic. Typical tricks for doing multiplication by a non-power-of-two constant would be to break the single multiplication into a series of shift-and-adds. So, multiplying by a constant "5" is the same as multiplying by 4 (left shift by 2), then adding itself.

Depending on how the multiplication is performed is going to affect system performance. Xilinx offers dedicated multipliers embedded in multiplication blocks or DSP48x modules. These hardware multipliers consume no "fabric" and typically run at or near the maximum clock rate of the silicon.

Unfortunately, division is another story. Certainly numeric methods have been developed for microprocessors and many of these techniques can be ported to the FPGA domain. Certainly the trivial case of dividing by a power-of-two can be implemented simply by right-shifting a signal. More complex division, such as fixed-point and floating-point math is available in the IEEE fractional packages (IEEE 1076.3-2006). Not all aspects of these packages may be efficiently implemented in synthesis.

5.7.4 *TEXTIO*

There are two different TEXTIO packages. Conveniently, they are intended to work together.

The first TEXTIO package is available in the STD library (which is always available), but this package must have an explicit "use" statement. Originally intended to support textio for the "built-in" data types in STD.STANDARD (which doesn't have to be explicitly loaded), it provides support for:

- bit/bit_vector
- integer, natural, positive
- Boolean
- character/string
- real
- time

And provides the following functions

- line – a pointer to a string which is the principle means of moving information in and out of a file
- text – a file type which can be mapped to a specific file on disk

- side – a type used in justification of output data
- width – used to specify the width of output data
- read/readline – read a data type from a line, read an entire line from a text file
- write/writeline – write a data type to a line, write an entire line to a text file
- endline – a Boolean function which is true when no more data can be read

These functions are not synthesizable and are used almost exclusively in simulation to read data sets and to log output data. While much can be said about the textio library functions, a short example should provide sufficient information to use on your own…

The second TEXTIO package is provided by the IEEE (IEEE.TEXTIO.all) which leverages and extends the capabilities of the STD.TEXTIO package to provide support for std_logic/std_logic_vector types and provide the ability to read std_logic_vectors as not only binary (READ/WRITE), but in octal (OREAD/OWRITE) and hexadecimal (HREAD/HWRITE).

```
 1: mkdata: process
 2:       variable bitVectorValue   : bit_vector(31 downto 0) :=
                                                      X"1234_5678";
 3:       variable bValue      : boolean := true;
 4:       variable strValue    : string(1 to 16) :=
                                           "important stuff!";
 5:       file dataFile        : TEXT is out "MY_DATA_FILE.TXT";
 6:       variable lineOfData : line;
 7:    begin
 8:      write(L => lineOfData, VALUE => strValue,
               JUSTIFIED => LEFT, FIELD => 20);
 9:      write(lineOfData,bValue,RIGHT);
10:      writeLine(dataFile,lineOfData);
11:      write(lineOfData,bitVectorValue, right, 64);
12:      writeLine(dataFile,lineOfData);
13:      wait for 10 ns;
14:      write(lineOfData,now);
15:      writeline(datafile,lineOfData);
16:      wait for 10 ns;
17:      wait;
18:   end process mkdata;
```

Snippet 5.39: Writing to a Text File Using Textio.

Code analysis:

Line 5: a file type of *dataFile* is defined as an output file with the name which will appear on disk as "MY_DATA_FILE.TXT".

Line 6: variable *lineOfData* is defined to be of type *line* which is a pointer[5] to a string. For now we will just use it to hold images of the data we want to write to the file.

Line 8: this write statement illustrates the use of named association when accessing procedures. The *write* procedure takes the pointer to the line of data to be written (which will be

[5] Pointers in VHDL are indicated by access types which can be quite useful when performing very complex simulations. They have been effectively omitted from this text as (1) most FPGA designers won't need them and presenting them here would only serve to defocus the reader and (2) it is a bit more of an advanced topic.

filled according to the other values), the VALUE parameter which accepts one of the types listed above as the value to add to the *lineOfData*, whether the value will be right or left justified within the field, and finally the width of the field. In this line, the value of the string "important stuff!" is pushed into the *lineOfData.*

Line 9: Next, the Boolean variable *bValue* is appended on to *lineOfData*. Notice that not all the fields are identified – fortunately, the WRITE procedure has a number of overloads so that not all data must be provided for each procedure call.

Line 10: the writeLine function takes the value currently stored in the string and moves it to the text file and appends a carriage return and line feed.

Line 11: this time *bitVectorValue* is placed into *lineOfData* in a 64-character right-justified line.

Line 12: this single piece of information is now moved out to the text file.

Line 14: the WRITE procedure supports the physical type of time. The *now* function returns the current time in the simulation environment.

Line 15: the text form of the time is now moved out to the text file (see Figure 5.17).

Now we can read our data back in using the conjugate function *read* and *readline:*

```
 1: rdData: process
 2:    variable bitVectorValue   : bit_vector(31 downto 0);
 3:    variable bValue           : boolean := true;
 4:    variable strValue         : string(1 to 20);
 5:    file dataFile             : TEXT is in "MY_DATA_FILE.TXT";
 6:    variable lineOfData       : line;
 7:    variable status           : boolean;
 8:    variable simTime          : time;
 9: begin
10:    if not endfile(dataFile) then
11:      readLine(dataFile,lineOfData);
12:      read(L=>lineOfData,VALUE=>strValue,GOOD=>status);
13:      report "strValue read back as: " & strValue;
14:      report "status is " & boolean'image(status);
15:      read(lineOfData,bValue);
16:
17:      readLine(dataFile,lineOfData);
18:      read(lineOfData,bitVectorValue);
19:      if (bitVectorValue = X"1234_5678") then
20:         report "bit vector matches hex 1234_5678";
21:      else
22:         report "bit vector failed to match";
23:      end if;
24:
25:      readLine(dataFile, lineOfData);
26:      read(lineOfData,simTime);
27:      report "simTime: " & time'image(simTime);
28:
29:    end if;
30:    wait;
31: end process rdData;
```

Snippet 5.40: Reading Back from the Data File.

```
important stuff!    TRUE

00010010001101000101011001111000
10 ns
```

Figure 5.17: The Output Data File.

Code analysis:

The std_logic_textio extends the data types supported for the *READ* and *WRITE* function to include std_logic and std_logic_vector. It provides the additional capability of reading and writing in hexadecimal (*HREAD* and *HWRITE*) and in octal (*OREAD* and *OWRITE*).

APPENDIX

A Quick Reference

A.1 Language Constructs

	Concurrent Statements		Sequential Statements

Unconditional assignment	target <= expression;
	z <= x OR y;
	mVec <= qVec and X"0F";
		↔	signal_target <= expression;
			z <= x OR y;
			
			mVec <= qVec and X"0F";
			
			variable_target: = expression;
			z: = x OR y;
			mVec: = qVec and X"0F";

Conditional assignment	target <= expr when condition else
	expr when condition else
	expr;
	
	z <= x when (q = '1') else
	y when (g = '0') else
	'0';
		↔	if (cond) then <statements>
			elsif (cond) then <statements>
			else <statements>
			end if;
			
			if (q = '1') then
			z <= x;
			elsif (g = '0') then
			z <= y;
			else z <= '0';
			end if;

Selective assignment	with selector select target <= expression when choice, expression when choice, ...; with (button) select response <= 'Y' when "01", 'N' when "10", 'X' when others;	↔	case selector is when choice => <statements>; when choice => <statements>; When others => <statements>; end case; case button is When "01" => response <= 'Y'; when "10" => response <= 'N'; when others => response <= 'X'; end case;

(continued on next page)

(continued)

	Concurrent Statements	Sequential Statements

Process statement	<lable>: process (sensitivity list)
	  begin
	  end process <label>;
	
	-- sync process with active high reset
	P1: process (clk)
	  begin
	    if rising_edge(clk) then
	      if (reset = '1') then
	      else
	      end if;
	    end if;
	  end process P1;
		***Sequential statements can only be used inside of processes, functions, and procedures ***

A.2 Data Types

A.3 Standard Libraries

The Standard Library contains two packages: standard package, which is always included with every Very-High-Speed Integrated Circuit Hardware Description Language (VHDL) file and does not need to be explicitly declared, and the textio package which does need a "use" clause to be accessed.

The standard package contains the data types for BOOLEAN (false,true), BIT ('0', '1'), CHARACTER (both printing and non-printing characters up to 8 bits), SEVERITY_ LEVEL (NOTE, WARNING, ERROR, FAILURE) which is used with assert statements, INTEGER which covers the minimum range of 32 bits (1 sign plus 2^31-1), TIME, REAL, STRING, as well as subtypes for INTEGER including NATURAL and POSITIVE.

The TEXTIO package extends the standard packages' capabilities for accessing text-based files. Included in this package are the definitions for type LINE and TEXT, as well as a number of overloaded functions for READ, READLINE, WRITE, WRITELINE, and ENDLINE. It is important to note that the functions defined in these packages do NOT support std_logic/std_logic_vector! These capabilities are included with the IEEE libraries described next.

Type	Values	Package
std_logic	'0', '1', 'Z', 'U', 'H', 'L', 'W', 'X', '-'	std_logic_1164
std_logic_vector	Array of std_logic	std_logic_1164
unsigned	Array of std_logic	numeric_std
signed	Array of std_logic	numeric_std
integer	$-(2^{31}-1)$ to $(2^{31}-1)$	standard
real	−1.0E38 to 1.0E38	standard
time	1 fs to 1 h	standard
Boolean	true, false	standard
character	256 characters as defined in	standard
String	Array of characters	standard

Index

Printed and bound by CPI Group (UK) Ltd, Croydon, CR0 4YY
03/10/2024
01040343-0001